Diagnosis and Surgical Treatment of Epilepsy

Diagnosis and Surgical Treatment of Epilepsy

Special Issue Editor

Warren W. Boling

MDPI • Basel • Beijing • Wuhan • Barcelona • Belgrade

MDPI

Special Issue Editor
Warren W. Boling
Loma Linda University Medical
Center
USA

Editorial Office
MDPI
St. Alban-Anlage 66
4052 Basel, Switzerland

This is a reprint of articles from the Special Issue published online in the open access journal *Brain Sciences* (ISSN 2076-3425) in 2018 (available at: http://www.mdpi.com/journal/brainsci/special_issues/epilepsy)

For citation purposes, cite each article independently as indicated on the article page online and as indicated below:

LastName, A.A.; LastName, B.B.; LastName, C.C. Article Title. *Journal Name* **Year**, *Article Number*, Page Range.

ISBN 978-3-03897-449-9 (Pbk)
ISBN 978-3-03897-450-5 (PDF)

Contents

About the Special Issue Editor

Warren W. Boling, MD, FAANS, FRCSC, FRACS, Professor and Chairman, Department of Neurosurgery, Loma Linda University. Dr. Boling completed his neurosurgery training at University of Kentucky. He then completed fellowships in Epilepsy and Functional Neurosurgery at Montreal Neurological Institute of McGill University, Canada and University of Melbourne, Australia. Dr. Boling began his career as Assistant Professor faculty in the Department Neurosurgery of McGill University and the MNI. He later served as Associate Professor of Neurosurgery at West Virginia University then Associate Professor Faculty, Department of Surgery at University of Melbourne. Dr. Boling was previously Professor of Neurosurgery and Interim Chairman of the Department of Neurosurgery, University of Louisville. Dr. Boling is currently Professor and Chairman of the Department of Neurosurgery at Loma Linda University. He is the past president of the Society of Brain Mapping and Therapeutics and a member of the SBMT Board of Directors. He is Board Certified by the American Board of Neurological Surgeons and a Fellow of the Royal College of Surgeons of Canada and the Royal Australasian College of Surgeons.

brain
sciences

MDPI

Editorial

Diagnosis and Surgical Treatment of Epilepsy

Warren Boling

MD, FAANS, FRCSC, FRACS, Loma Linda University Medical Center, Loma Linda, CA 92354, USA;
wboling@LLU.edu; Tel.: +1-909-558-4419

Received: 13 June 2018; Accepted: 14 June 2018; Published: 21 June 2018

Epilepsy is a common neurological disease that can affect all ages. Although the majority of people with epilepsy will have excellent seizure control with medication, about 30% will fail anti-epileptic drugs. For those with medically intractable epilepsy, the recurrent seizures lead to increased mortality, risks of injury, and the seizures themselves are socially disabling [1–3]. Fortunately for many people with intractable epilepsy, epilepsy can be cured, or seizures better controlled with surgical treatment [4,5].

Localization of the seizure focus followed by surgical resection provides the best opportunity to cure the epilepsy, and a better understanding of the neuro-anatomy and physiology of epilepsy improves our ability to define the epileptic network and effectively treat the epilepsy [6]. An important advance that has improved patient care is minimal access surgical approaches, which result in more rapid recovery from surgery, less pain, and more satisfied patients [7].

Additionally, for individuals without an opportunity for cure of their epilepsy, new and emerging technologies have promise to reduce seizure frequency and severity, thus improving quality of life and preventing injuries and mortality that result from intractable epilepsy.

In this special issue, Boling et al. examines the profound negative consequences of medically intractable epilepsy that impacts the majority of the world's population who reside in the developing world of Asia and sub-Saharan Africa. Stigma is a major driver of the significantly reduced quality of life of people with epilepsy, which is amplified in severely underserved and low-resource regions of the world due to poverty, severe treatment gaps, high mortality and morbidity of intractable epilepsy, lack of education and knowledge about epilepsy, and widespread misconceptions that epilepsy is related to witchcraft or sorcery as well as beliefs that epilepsy is contagious. Boling et al. then describe proof of principle in the developing world that surgery of medically intractable epilepsy can elevate quality of life and significantly reduce stigma.

Even in the developed world, due to seemingly inextricable reasons, wait times for patients with medically intractable epilepsy to be evaluated in a comprehensive epilepsy program are unnecessarily long. The delay to evaluation of surgically remedial epilepsies results in many patients being exposed far too long to the elevated mortality risk and reduced quality of life that results from intractable seizures. Sadanand explores this knotty problem using a novel mathematical approach of non-cooperative game theory. He then contrasts and compares the medical communities approach to glioblastoma multiforme, which has better defined treatment algorithms and expectations of care, with the medical community's approach to intractable epilepsy treatment, and explains the discrepancies identified using game theory models.

Anyanwu et al. reviews the definition of medical intractability and provides a broad overview of the treatment options available to patients who have failed medication. Although approximately 20 anticonvulsant medications are available in North America and Europe today, the authors point out that only two anticonvulsant medications need to be adequately trialed and failed prior to a patient being deemed intractable. The presurgical evaluation particularly EEG and imaging are discussed. Finally the various surgical procedures for both palliative and curative goals are covered.

New treatment options have recently become available for epilepsy patients that directly stimulate the brain to suppress the seizure focus. The Neuropace RNS system (Mountain View, CA, USA) is a closed loop device that monitors EEG in order to identify a seizure onset then delivers a stimulus to prevent the seizure from spreading to become symptomatic. Most recently, the FDA in the United States approved anterior nucleus of the thalamus stimulation for adults with medically intractable partial epilepsy using a Medtronic DBS stimulation device (Minneapolis, MN, USA). Two articles, one from Kwon et al. and another from Eastin et al. explore the recent developments and historical underpinnings of neuromodulation treatments of epilepsy. Eastin et al. reviews the various stimulation targets that have been explored for neuromodulation of epilepsy, and they discuss many of the individuals who have championed these efforts. Kwon et al. discuss mechanisms of action in neuromodulation treatment of epilepsy then the authors specifically explore randomized controlled trials of stimulation treatment strategies for epilepsy that have been published in the literature. The pioneering work of Irving Cooper in the 1970s was the first human brain stimulation performed for epilepsy, but randomized studies that followed soon after did not show benefit with a cerebellar target. Most of the modern interest in brain neuromodulation has focused on thalamic and supratentorial cortical targets, plus there is promising research targeting the hippocampus for stimulation. However, the most common neuromodulation strategy for epilepsy today continues to be vagus nerve stimulation.

In this special issue, Boling describes the various surgical approaches, nuances and pitfalls of surgery of medically intractable temporal lobe epilepsy with an emphasis on mesial temporal lobe epilepsy (MTLE). Selective and keyhole approaches to treat intractable MTLE allow for more rapid recovery of the patient. Recently after two new laser ablation devices became available there has been a resurgent interest in thermal ablation as a treatment option for MTLE. Despite the advances of minimal access keyhole surgery and thermal ablation techniques, an accurate diagnosis of MTLE remains paramount in the surgical treatment success of medically intractable MTLE.

High quality imaging is a critical component of the evaluation of intractable epilepsy. Identification of a lesion aides in defining the epileptogenic zone and significantly improves the seizure freedom opportunity of surgery. Skull base temporal lobe encephaloceles are fascinating and often overlooked lesions that can result in temporal lobe epilepsy. Bannout et al. examined their groups experience with these lesions and reviewed the literature in order to determine if a limited lesionectomy of the encephalocele was sufficient or a more extensive anterior temporal lobe resection was required to achieve seizure freedom in cases of medical intractability.

Finally, Hussein et al. evaluates in an animal model a common nutritional supplement L-Carnitine that seems to have an anticonvulsant effect. The authors identify that the current antiepileptic medications are ineffective in about 30% of people with epilepsy who will be intractable, and anticonvulsants used today mostly reduce neuronal excitation in order to lower the seizure threshold. In a rat model of epilepsy, the authors identified that L-Carnitine was associated with a marked reduction of the seizure frequency and shortened the seizure duration. Also, rats treated with L-Carnitine had relative neuroprotection with less neuronal death identified in the hippocampus. The beneficial effects of L-Carnitine were demonstrated by Hussein et al. to work through a novel mechanism of reduction of oxidative stress and up regulation of heat shock proteins. The authors caution that further research needs to be done to further elucidate mechanisms of action. However, L-Carnitine is promising as a novel class of anticonvulsant medication with a mechanism of action different from the standard anticonvulsants that lower the threshold of neuronal excitability.

Conflicts of Interest: The author declares no conflict of interest.

References

1. Cockerell, O.C.; Johnson, A.L.; Sander, J.W.A.S.; Hart, Y.M.; Goodridge, D.M.G.; Shorvon, S.D. Mortality from epilepsy: Results from a prospective population-based study. *Lancet* **1994**, *344*, 918–921. [CrossRef]
2. Sperling, M.R.; Feldman, H.; Kinman, J.; Liporace, J.D.; O'Connor, M.J. Seizure control and mortality in epilepsy. *Ann. Neurol.* **1999**, *46*, 45–50. [CrossRef]

3.   Baker, G.A. The psychosocial burden of epilepsy. *Epilepsia* **2002**, *43*, 26–30. [CrossRef] [PubMed]
4.   Kwan, P.; Arzimanoglou, A.; Berg, A.T.; Brodie, M.J.; Hauser, W.A.; Mathern, G.; Moshé, S.L.; Perucca, E.; Wiebe, S.; French, J. Definition of drug resistant epilepsy. Consensus proposal by the ad hoc Task Force of the ILAE Commission on Therapeutic Strategies. *Epilepsia* **2010**, *51*, 1069–1077. [CrossRef] [PubMed]
5.   Fletcher, A.; Sims-Williams, H.; Wabulya, A.; Boling, W. Stigma and quality of life at long-term follow-up after surgery for epilepsy in Uganda. *Epilepsy Behav.* **2015**, *52*, 128–131. [CrossRef] [PubMed]
6.   Olivier, A.; Boling, W.; Tanriverdi, T. *Techniques in Epilepsy Surgery, The MNI Approach*, 1st ed.; Cambridge University Press: Cambridge, UK, 2012.
7.   Boling, W. Minimal access keyhole surgery for mesial temporal lobe epilepsy. *J. Clin. Neurosci.* **2010**, *17*, 1180–1184. [CrossRef] [PubMed]

Review

Stimulation and Neuromodulation in the Treatment of Epilepsy

Timothy Marc Eastin and Miguel Angel Lopez-Gonzalez *

Department of Neurosurgery, Loma Linda University, Loma Linda, CA 92354, USA; meastin@llu.edu
* Correspondence: mlopezgonzalez@llu.edu; Tel.: +1-909-558-0800

Received: 3 November 2017; Accepted: 18 December 2017; Published: 21 December 2017

Abstract: Invasive brain stimulation technologies are allowing the improvement of multiple neurological diseases that were non-manageable in the past. Nowadays, this technology is widely used for movement disorders and is undergoing multiple clinical and basic science research for development of new applications. Epilepsy is one of the conditions that can benefit from these emerging technologies. The objective of this manuscript is to review literature about historical background, current principles and outcomes of available modalities of neuromodulation and deep brain stimulation in epilepsy patients.

Keywords: brain stimulation; neuromodulation; epilepsy; surgery

1. Introduction

It is estimated that epilepsy affects more than 50 million people worldwide. The majority of those patients (60–70%) reach an adequate control based on pharmacological treatments, while a refractory group of patients can become candidates for epilepsy surgery. In some situations, surgical resection is not feasible due to expected increased morbidity for epileptogenic focus located within highly functional cortical areas. In temporal lobe epilepsy, 30–40% of patients do not improve satisfactorily after resective surgery [1]. The failure rate may increase among reported series due to multiple factors, such as location of surgical target (frontal lobe, insula), classification of failure according to studies (strict Engel IV classification, or both Engel III and IV classification).

Neuromodulation allows the possibility to treat different pathologies as reversible and non-lesional alternatives. The term "neuromodulation" is essentially electrical stimulation of the nervous system in order to modulate or modify a specific function (as in movement disorders, pain, epilepsy), and can be delivered in different ways: through stimulation over skin surface, peripheral nerve stimulation, cortical stimulation, or deep brain stimulation. The different neuromodulation systems are connected to an implantable pulse generator where different stimulation settings can be modified (Figure 1). Different locations and targets for surgically implanted neuromodulation systems have been proposed for the management of epilepsy, and this review will analyze the success rate and outcomes within the available literature.

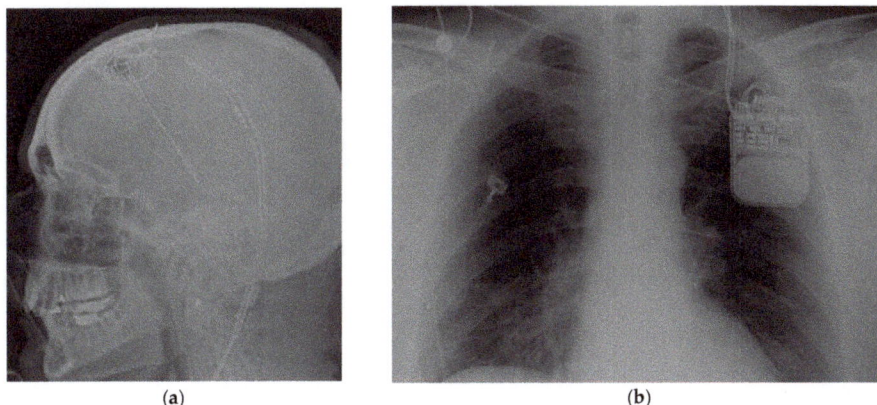

(a)   (b)

Figure 1. (**a**) Lateral skull X-rays view of thalamic deep brain stimulator system; (**b**) Left infraclavicular implantable pulse generator.

2. History of Neuromodulation in Epilepsy

The initial evidence of neuromodulation dates from 15 CE (Common Era) by Scribonius; he observed that gout pain improved by accidental contact with a torpedo fish and after this discovery, electrical shock was used to treat multiple types of pain [2]. Sir Victor Horsley, in 1886, performed the first cortical stimulation for focal lesion resection in patient with epilepsy; by 1908, while he was associated with Clark, they developed the stereotaxis frame for experimental stimulation in the lab [3]. Fedor Krause, between 1893 and 1912, continued performing cortical stimulation, producing the first accurate map of human motor strip, later improved by Foerster, a neurologist trained by Dejerine and Wernicke. Foerster later applied his neurological anatomical knowledge to perform surgical procedures [4]. Wilder Penfield learned the motor mapping techniques from Foerster in Germany. He eventually went beyond with mapping of speech, hearing, vision, and memory functions [4]. After 1948, the year when Spiegel and Wycis published their experience on human stereotaxy, the surgical activity in the stereotactic field expanded dramatically, with the development of multiple other frames and techniques [5]. The experimental models used by Hassler in the 1940s allowed him to develop the thalamic atlas with the nomenclature most widely used nowadays [6]. During the 1950s and 1960s the stereotactic lesions were commonly used in epilepsy with the evidence of change in electroencephalographic recordings during stimulation of thalamus and globus pallidus [2]. In 1955, the cerebellar cortex stimulation was studied by Cooke and Snider [7], and the first trial for chronic cerebellar stimulation that suggested decreased seizure frequency was performed by Cooper in 1973 [8]. This was followed by multiple studies for deep cerebellar [9], and thalamic centromedian nucleus stimulation [10]. The anterior thalamic nucleus stimulation was analyzed with SANTE trial (Stimulation of anterior nucleus of the thalamus in epilepsy) although not obtaining Food and Drug Administration (FDA) approval [11]. Other potential central nervous system (CNS) targets such as the hippocampus [12] and subthalamic nucleus [13] were analyzed, evolving more recently to the direct cortical stimulation or responsive neurostimulation [14,15] which was FDA-approved in 2013.

Peripheral nervous system stimulation in epilepsy was initially performed by Bailey in 1938. He developed a cat model for afferent vagal stimulation that modified electroencephalogram activity [16]. The mechanism described involved the effect in locus ceruleus and nucleus solitarius, and its initial clinical application of vagal nerve stimulation was performed in 1988 [17], obtaining FDA approval in 1997. More recently, based on hypothesis of connections between locus ceruleus, nucleus solitarius and trigeminal nucleus, the idea of external trigeminal nerve stimulation through transdermal or subcutaneous electrodes to stimulate trigeminal nerve supraorbital branches was triggered. This is not FDA approved yet [18].

3. Central Nervous System Stimulation

3.1. Cerebellar Stimulation

This was initially performed by Cooper in 1973, based on the idea of inhibitory effect over efferent pathways. He reported a 50% reduction in seizure activity in 18 out of 32 patients [8]. Experimental studies favored upper medial cerebellar cortex stimulation as a potential target for generalized seizures, while deep cerebellar nuclei could potentially control limbic seizures, although other studies did not show any effect at all [19]. Van Buren et al. [9], in a case series did not show seizure improvement, further contradicted by Velasco et al. [20], but later, a systematic review showed inconsistent outcomes [21].

3.2. Thalamic Stimulation

Thalamic nucleus stimulation is well established for treatment of movement disorders, such as essential tremor, with FDA approval since 1997. The initial ideas of therapeutic thalamic involvement on epileptic activity dates back from stereotactic lesions for seizure control since the 1960s [22], while pioneering deep brain stimulation for epilepsy was reported by Cooper and Upton until 1985 [23]. Two specific targets have been studied, the anterior thalamic nucleus and centromedian thalamic nucleus.

3.2.1. Anterior Thalamic Nucleus

The rationale for anterior thalamic nucleus (AN) stimulation is based on its role as a primary relay nucleus of the limbic system, receiving projections from mammillary bodies, cingulum, amygdala, hippocampus and orbito-frontal cortex [24,25]. Isolated case series, with seizure suppression and acceptable safety features, prompted a larger multicenter trial, the SANTE study (stimulation of the anterior nucleus of thalamus for epilepsy) [11]. A total of 110 patients in 17 US institutions were included, and showed 40% reduction in seizure frequency in the treatment group against 15% reduction in the control group during the blinded on/off evaluation. At a 12-month open label follow-up, 54% of patients showed more than 50% seizure reduction. The complications reported included paresthesias (18.2%), infection (12.7%), hemorrhage (4.5%), status epilepticus (4.5%) and death (4.5%). The deaths were attributed to sudden unexpected death in epilepsy (SUDEP), and did not occur within 30 days after surgery. It is considered that AN deep brain stimulation can benefit frontal–temporal onset epilepsies. This is currently approved in Europe and Canada, but not FDA approved.

3.2.2. Centromedian Thalamic Nucleus

The Centromedian (CM) thalamic nucleus, initially described by Luys in 1895, is considered part of the intralaminar nucleus of the thalamus, and has been associated with many physiological and pathological states [26]. Wilder and Penfield first postulated that this nucleus could modulate or ameliorate seizures. They receive extensive input from mesencephalic, pontine, and medullary reticular formation, while the output is described as diffuse and non-specific. Stimulation in these areas causes the so-called cortical recruiting effect. Low-frequency stimulation causes slow-wave electroencephalogram (EEG) activity associated with somnolence, while high-frequency stimulation results in desynchronized cortical activity, arousal and even epileptiform activity if stimulation is very high [27]. Extensive research and clinical work by Velasco et al., reported an overall benefit of CM stimulation with seizure frequency reduction of generalized tonic clonic seizures by 80–100%, and 60% for complex partial seizures [28]. These results were similar to those reported by Fisher et al. [29] and Valentin et al. [30], with more than 50% reduction of generalized seizure frequency. Overall, these results derived from case series with small numbers of patients. With current information, it appears that CM nucleus stimulation can be considered as a treatment option in refractory generalized epilepsy, although stimulation is less effective in frontal lobe epilepsies. At present, this target stimulation is not FDA approved.

3.3. Mesiotemporal Stimulation

The role of amygdala and hippocampus in epilepsy has been long studied in experimental models, and also from clinical experience, with surgical resections and success over medically intractable epilepsy. The hypothesis of hippocampal stimulation for epilepsy is based on in vitro studies exploring the effect of electrical stimulation showing suppression of epileptiform discharges with different forms of stimulation [31]. In experimental models, low-frequency stimulation inhibits the development and progression of seizures, while high-frequency stimulation increases threshold and latency of after-discharges [32]. Current use of responsive stimulation is giving clinical information about high frequency stimulation parameters and its efficacy. This option of neuromodulation has been considered when bilateral onset of mesial temporal epilepsy is present, or when high risk of neurocognitive deficit is expected after surgical resection.

Few clinical studies with small numbers of patients were conducted, showing overall, more than 50% seizure reduction. The initial study of hippocampal stimulation in epilepsy was conducted by Velasco et al. [12] in 2000, studying ten patients with a short follow-up before temporal lobectomy. The potential mechanism of action proposed was activation of perforant pathway with inhibitory influence on epileptogenic neurons of areas cornu ammonis 1–4 (CA1–CA4). They found that in 70% of patients, seizures were abolished. Later, the same group performed a randomized controlled trial with longer follow-up in nine patients with bitemporal seizure onset, and encountered seizure reduction of >95% in five patients with normal magnetic resonance imaging (MRI) and between 50–70% seizure reduction in four patients with presence of mesial temporal sclerosis [33]. Neuropsychological testing in those patients did not show impairment after stimulation period. Another randomized controlled trial by Tellez-Zenteno et al. [1] in 2006, studied four patients with unilateral mesial temporal epilepsy and hippocampal sclerosis that had high risk for memory loss with resective surgery. The reduction in seizure frequency was 15% during "On" periods. The comparison with seizure frequency at baseline showed a median reduction of 26% when the stimulator was "On", and an increase of 49% when the stimulator was "Off". Given the small number of patients in all available studies with mesial temporal stimulation, there is no robust evidence for its wide application. This option can be considered in patients with high risk of memory or neuropsychological decline after resective surgery, and currently, widespread use of responsive neurostimulation will allow for gathering of more data regarding its effectiveness.

3.4. Subthalamic Nucleus Stimulation

Subthalamic nucleus (STN) is a well-documented target for its clinical efficacy in medically intractable Parkinson's disease either using frame or frameless stimulation techniques (Figure 2). The effect of stimulation for epilepsy in experimental models suggested some efficacy on seizure control. The proposed mechanism is by increasing glutamate input from STN, increasing gamma-Aminobutyric acid GABA-ergic firing from substantia nigra pars reticulata (SNr), and later, causing inhibition over the dorsal midbrain antiepileptic zone [34]. The first clinical application of STN stimulation in epilepsy was performed by Benabid et al. [13] in 2002 in one patient, showing 80% reduction in seizures for more than two years following treatment. Subsequent reports favored inferior subthalamic nucleus stimulation, but unfortunately, only included a very small number of patients and lacking of solid results. STN is also considered an experimental target in epilepsy, and will require controlled randomized trials and detailed neuropsychological evaluations in order to evaluate efficacy and safety.

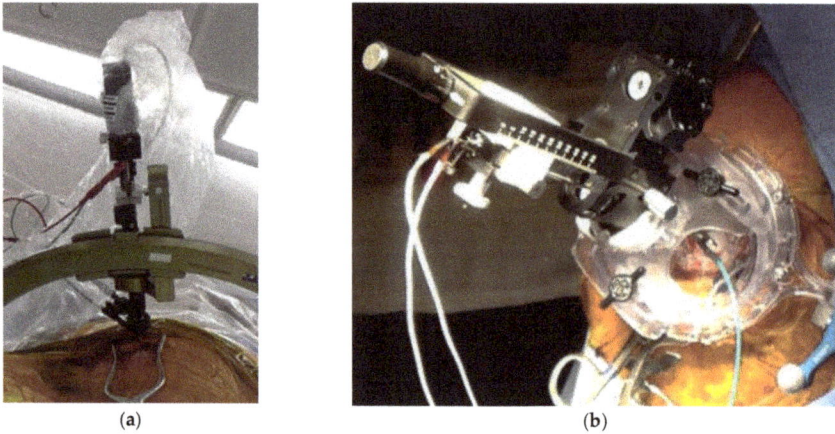

(a)   (b)

Figure 2. (**a**) Frame based system for deep brain stimulation; (**b**) Frameless bases system for deep brain stimulation.

3.5. Cortical Stimulation

Chronic subdural cortical stimulation: The use of surgical electrocorticography during functional brain mapping and lesion resection trigger after discharges that can progress to full seizures. Brief pulses of stimulation initially performed by Lesser et al. [35] were able to control after discharges; the mechanism of this effect is not clearly understood. In theory, stimulation trigger alterations to GABA, calcium channels, and extracellular potassium, that can induce depolarization. Experimental and small case reports trying to optimize stimulation parameters are still under investigation to clarify its application [36,37].

Responsive stimulation: The principle of responsive neurostimulation is the first closed-loop system available with the ability of both recording and stimulating. This system involves implantation of subdural and depth electrodes connected to a skull implant. The electrodes are able to register minutes-worth of data from sliding-window of EEG recordings, sense seizure onset and to deliver electrical activity with the objective to abort and inhibit the spread of epileptogenic activity (Figure 3). The device is intended to be placed in eloquent areas, and the initial clinical experience was by Osorio et al. [15], and after further studies, it was approved by FDA in 2013. A large multicenter trial by Morrell et al. [38] evaluated efficacy with seizure reduction of 44% from baseline at one year, and 65.7% by six years. The main complication rate encountered was implant site infection in (9.4%), and hemorrhage (4.7%). Half of the patients had temporal lobe epilepsy, and among them, the majority had bitemporal onset. The ability to sense epileptic activity, quantify frequency of seizures, seizure duration activity, preceding electroencephalographic events that vary among patients, and to deliver immediate electrical stimulation, are significant advantages over other neuromodulation systems, and can open the gateway for development and improvement of the technique.

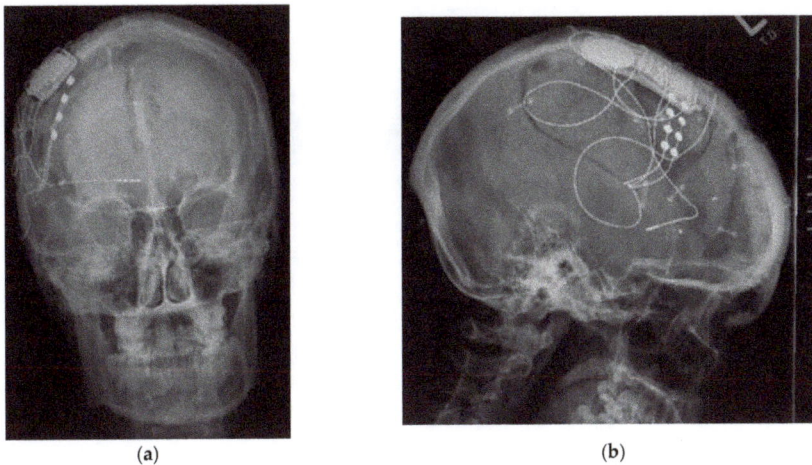

(a)   (b)

Figure 3. (a) Antero-posterior skull X-rays of implanted responsive neurostimulation; (b) Lateral skull X-rays of implanted responsive neurostimulation.

4. Peripheral Nervous System Stimulation

4.1. Vagal Nerve Stimulation (VNS)

The inhibition of motor activity by stimulation of vagal nerve afferents was reported by Schweitzer and Wright in 1937 [39]. In 1938, a cat model was developed, where afferent vagal stimulation modified electroencephalogram activity [16]. The exact mechanism of vagal nerve stimulation to control seizures is not well understood; however, the hypothesis is based on the anatomic projections of the vagus nerve to reticular activating system, central autonomic network, limbic system and noradrenergic projection system [17,40].

It is estimated that about 80% of the vagus nerve is composed of afferent fibers arising from viscera, while around 20% is composed of efferent fibers that innervate larynx and have parasympathetic control of the heart, lungs, and gastrointestinal system [41]. The main trunk of the vagus nerve is easily exposed, surgically, within the carotid sheath between and behind the internal jugular vein and common carotid artery (Figure 4). The efferent cardiac branches join the sympathetic fibers high within the neck, and are very unlikely to reach those branches during standard dissection. The right vagal nerve innervates the sinoatrial node, and for that reason, the vagal nerve stimulator is implanted on the left side. Overall, given anatomical variability, it is important to communicate with anesthesia team at the time of initial stimulation, in order to manage remote, but possible, bradycardia and asystole.

The vagal nerve stimulation is indicated for medically refractory partial onset seizures, although commonly used for generalized seizures [42]. The largest long-term study by Morris et al. [43] showed a median seizure reduction of 35% at one year and 44% at three years. Interestingly, another smaller long-term study showed 65.7% seizure reduction at six years after implantation, and 75.7% seizure reduction at 10 years [44]. It was FDA approved since 1997, and the new model has a closed-loop system that can detect a 20% or higher sudden increase in heart rate, considering that this often signals the onset of a seizure. This new model was FDA approved in 2015. The adverse events reported are hoarseness and cough in around 50% of patients, which can later be improved by habituation or modification of stimulation settings. By two years of stimulation, reported hoarseness decreased to 19.8% [43]. Serious complications are uncommon, such as vocal cord paralysis (1%) and infection (1.5%) [45]. Recent studies show that earlier use of VNS therapy is proven to offer better long-term outcomes for children [46] which favored FDA approval in 2017 for children as young as four years old.

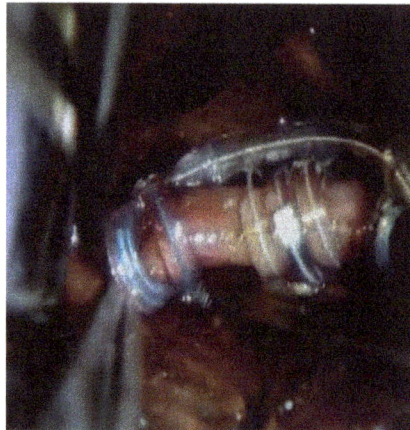

Figure 4. Vagus nerve stimulator placement.

4.2. Trigeminal Nerve Stimulation

In 1976, Maksimow et al. [47] reported the interruption of seizures by the trigeminal nerve stimulation. The current device available in Europe and Canada is non-invasive, and stimulates the supraorbital and infraorbital nerves with transdermal electrodes. The rationale for its use is similar to that of VNS, theoretically generating stimulation transmitted to the reticular activating system. In a Phase II trial by Soss et al. [48], the seizure control was reported to be 27.4% at six months and 34.8% at 12 months. The bilateral stimulation is favored over unilateral stimulation.

5. Conclusions

The combination of pharmacological resistance in epilepsy, exhaustion, or inability to perform safe surgical resective options in some severe epilepsy cases open the highway for neuromodulation in epilepsy. There is a need for coupling the historical background of neuromodulation research with recent technological and computerized advances for a better understanding of available targets. The horizon of these technologies is huge. It is imperative to perform formal and large-scale clinical trials to give solid indications of neuromodulation and deep brain stimulation techniques in epilepsy.

Author Contributions: Miguel Angel Lopez-Gonzalez conceived, designed and wrote the manuscript. Timothy Marc Eastin wrote the manuscript.

Conflicts of Interest: The authors declare no conflict of interest.

References

1. Tellez-Zenteno, J.F.; McLachlan, R.S.; Parrent, A.; Kubu, C.S.; Wiebe, S. Hippocampal electrical stimulation in mesial temporal lobe epilepsy. *Neurology* **2006**, *66*, 1490–1494. [CrossRef] [PubMed]
2. Gildenberg, P.L. History of electrical neuromodulation for chronic pain. *Pain Med.* **2006**, *7*, S7–S13. [CrossRef]
3. Vilensky, J.A.; Gilman, S. Horsley was the first to use electrical stimulation of the human cerebral cortex intraoperatively. *Surg. Neurol.* **2002**, *58*, 425–426. [CrossRef]
4. Feindel, W.; Leblanc, R.; Villemure, J.G. History of the surgical treatment of epilepsy. In *A history of Neurosurgery in Its Scientific and Professional Contexts*; Greenblatt, S.H., Dagi, T.F., Epstein, M.H., Eds.; American Association of Neurological Surgeons: Rolling Meadows, IL, USA, 1997; pp. 465–488, ISBN 1-879284-17-0.
5. Spiegel, E.A.; Wycis, H.T.; Marks, M. Stereotaxic apparatus for operations on the human brain. *Science* **1947**, *106*, 349–350. [CrossRef] [PubMed]

6. Krack, P.; Dostrovsky, J.; Ilinksy, I.; Kultas-Ilinsky, K.; Lenz, F.; Lozano, A.; Vitek, J. Surgery of the motor thalamus: Problems with the present nomenclatures. *Mov. Disord.* **2002**, *17*, S2–S8. [CrossRef] [PubMed]

7. Cooke, P.M.; Snider, R.S. Some cerebellar influences on electrically-induced cerebral seizures. *Epilepsia* **1955**, *4*, 19–28. [CrossRef] [PubMed]

8. Cooper, I.S.; Amin, I.; Gilman, S. The effect of chronic cerebellar stimulation upon epilepsy in man. *Trans. Am. Neurol. Assoc.* **1973**, *98*, 192–196. [PubMed]

9. Van Buren, J.M.; Wood, J.H.; Oakley, J.; Hambrecht, F. Preliminary evaluation of cerebellar stimulation by double-blind stimulation and biological criteria in the treatment of epilepsy. *J. Neurosurg.* **1978**, *48*, 407–416. [CrossRef] [PubMed]

10. Velasco, F.; Velasco, M.; Ogarrio, C.; Fanghanel, G. Electrical stimulation of the centromedian thalamic nucleus in the treatment of convulsive seizures: A preliminary report. *Epilepsia* **1987**, *28*, 421–430. [CrossRef] [PubMed]

11. Fisher, R.; Salanova, V.; Witt, T.; Worth, R.; Henry, T.; Gross, R.; Oommen, K.; Osorio, I.; Nazzaro, J.; Labar, D.; et al. Electrical stimulation of the anterior nucleus of thalamus for treatment of refractory epilepsy. *Epilepsia* **2010**, *51*, 899–908. [CrossRef] [PubMed]

12. Velasco, M.; Velasco, F.; Velasco, A.L.; Boleaga, B.; Jimenez, F.; Brito, F.; Marquez, I. Subacute electrical stimulation of the hippocampus blocks intractable temporal lobe seizures and paroxysmal EEG activities. *Epilepsia* **2000**, *41*, 158–169. [CrossRef] [PubMed]

13. Benabid, A.L.; Minotti, L.; Koudsie, A.; de Saint Martin, A.; Hirsch, E. Antiepileptic effect of high-frequency stimulation of the subthalamic nucleus (corpus luysi) in a case of medically intractable epilepsy caused by focal dysplasia: A 30 month follow up: Technical case report. *Neurosurgery* **2002**, *50*, 1385–1391. [PubMed]

14. Osorio, I.; Frei, M.G.; Wilkinson, S.B. Real-time automated detection and quantitative analysis of seizures and short-term prediction of clinical onset. *Epilepsia* **1998**, *39*, 615–627. [CrossRef] [PubMed]

15. Osorio, I.; Frei, M.G.; Sunderam, S.; Giftakis, J.; Bhavaraju, N.C.; Schaffner, S.F.; Wilkinson, S.B. Automated seizure abatement in humans using electrical stimulation. *Ann. Neurol.* **2005**, *57*, 258–268. [CrossRef] [PubMed]

16. Bailey, P.B.F. A sensory cortical representation of the vagus nerve. *J. Neurophysiol.* **1938**, *1*, 405–412.

17. Penry, J.K.; Dean, J.C. Prevention of intractable partial partial seizures by intermittent vagal stimulation in humans: Preliminary results. *Epilepsia* **1990**, *31*, s40–s43. [CrossRef] [PubMed]

18. DeGiorgio, C.; Murray, D.; Markovic, D.; Whitehurst, T. Trigeminal nerve stimulation for epilepsy: Long-term feasibility and efficacy. *Neurology* **2009**, *72*, 936–938. [CrossRef] [PubMed]

19. Lockard, J.S.; Ojemann, G.A.; Congdon, W.C.; DuCharme, L.L. Cerebellar stimulation in alumina-gel monkey model: Inverse relationship between clinical seziures and EEG interictal burst. *Epilepsia* **1979**, *20*, 223–234. [CrossRef] [PubMed]

20. Velasco, F.; Carrillo-Ruiz, J.D.; Brito, F.; Velasco, M.; Velasco, A.L.; Marquez, I.; Davis, R. Double-blind, randomized controlled pilot study of bilateral cerebellar stimulation for treatment of intractable motor seizures. *Epilepsia* **2005**, *46*, 1071–1081. [CrossRef] [PubMed]

21. Fountas, K.N.; Kapsalaki, E.; Hadjigeorgiou, G. Cerebellar stimulation in the management of medically intractable epilepsy: A systematic and critical review. *Neurosurg. Focus* **2010**, *29*, E8. [CrossRef] [PubMed]

22. Mullan, S.; Vailati, G.; Karasick, J.; Mailis, M. Thalamic lesions for the control of epilepsy. A study of nine cases. *Arch. Neurol.* **1967**, *16*, 277–285. [CrossRef] [PubMed]

23. Cooper, I.S.; Upton, A.R.M. Therapeutic implications of modulation of metabolism and functional activity of cerebral cortex by chronic stimulation of cerebellum and thalamus. *Biol. Psychiatry* **1985**, *20*, 809–811. [CrossRef]

24. Mirski, M.A.; Rossell, L.A.; Terry, J.B.; Fisher, R.S. Anticonvulsant effect of anterior thalamic high frequency electrical stimulation in the rat. *Epilepsy Res.* **1997**, *28*, 89–100. [CrossRef]

25. Hamani, C.; Hodaie, M.; Chiang, J.; del Campo, M.; Andrade, D.M.; Sherman, D.; Mirski, M.; Mello, L.E.; Lozano, A.M. Deep brain stimulation of the anterior nucleus of the thalamus: Effects of electrical stimulation on pilocarpine-induced seizures and status epilepticus. *Epilepsy Res.* **2008**, *78*, 117–123. [CrossRef] [PubMed]

26. Fisher, R.S. Deep brain stimulation for epilepsy. In *Handbook of Clinical Neurology*; Lozano, A.M., Hallett, M., Eds.; Elsevier: Amsterdam, Netherlands, 2013; Volume 116, pp. 217–234, ISBN 978-0-444-53497-2.

27. Van der Werf, Y.D.; Witter, M.P.; Groenewegen, H.J. The intralaminar and midline nuclei of the thalamus. Anatomical and functional evidence for participation in processes of arousal and awareness. *Brain Res. Rev.* **2002**, *39*, 107–140. [CrossRef]

28. Velasco, F.; Velasco, A.L.; Velasco, M.; Jimenez, F.; Carrillo-Ruiz, J.D.; Castro, G. Centromedian thalamic stimulation for epilepsy. In *Textbook of Stereotactic and Functional Neurosurgery*, 2nd ed.; Lozano, A.M., Gildenberg, P.L., Tasker, R.R., Eds.; Springer: Berlin/Heidelberg, Germany, 2009; Volume 2, pp. 2777–2791, ISBN 978-3-540-70779-0.

29. Fisher, R.S.; Uematsu, S.; Krauss, G.L.; Cysyk, B.J.; McPherson, R.; Lesser, R.P.; Gordon, B.; Schwerdt, P.; Rise, M. Placebo-controlled pilot study of centromedian thalamic stimulation in treatment of intractable seizures. *Epilepsia* **1992**, *33*, 841–851. [CrossRef] [PubMed]

30. Valentin, A.; Garcia Navarrete, E.; Chelvarajah, R.; Torres, C.; Navas, M.; Vico, L.; Torres, N.; Pastor, J.; Selway, R.; Sola, R.G.; et al. Deep brain stimulation of the centromedian thalamic nucleus for the treatment of generalized and frontal epilepsies. *Epilepsia* **2013**, *54*, 1823–1833. [CrossRef] [PubMed]

31. Tellez-Zenteno, J.F.; Wiebe, S. Hippocampal stimulation in the treatment of epilepsy. *Neurosurg. Clin. N. Am.* **2011**, *22*, 465–475. [CrossRef] [PubMed]

32. Wyckhuys, T.; De Smedt, T.; Claeys, P.; Raedt, R.; Waterschoot, L.; Vonck, K.; Van den Broecke, C.; Mabilde, C.; Leybaert, L.; Wadman, W.; et al. High frecuency deep brain stimulation in the hippocampus modifies seizure charactheristics in kindled rats. *Epilepsia* **2007**, *48*, 1543–1550. [CrossRef] [PubMed]

33. Velasco, A.L.; Velasco, F.; Velasco, M.; Trejo, D.; Castro, G.; Carrillo-Ruiz, J.D. Electrical stimulation of the hippocampal epileptic foci for seizure control: A double-blind, long-term follow up study. *Epilepsia* **2007**, *48*, 1895–1903. [CrossRef] [PubMed]

34. Vercueil, L.; Benazzouz, A.; Deransart, C.; Bressand, K.; Marescaux, C.; Depaulis, A.; Benabid, A.L. High-frequency stimulation of the subthalamic nucleus suppresses absence seizures in the rat: Comparison with neurotoxic lesions. *Epilepsy Res.* **1998**, *31*, 39–46. [CrossRef]

35. Lesser, R.P.; Kim, S.H.; Beyderman, L.; Miglioretti, D.L.; Webber, W.R.; Bare, M.; Cysyk, B.; Krauss, G.; Gordon, B. Brief bursts of pulse stimulation terminate afterdischarges caused by cortical stimulation. *Neurology* **1999**, *53*, 2073–2081. [CrossRef] [PubMed]

36. Child, N.D.; Stead, M.; Wirrell, E.C.; Nickels, K.C.; Wetjen, N.M.; Lee, K.H.; Klassen, B.T. Chronic subthreshold subdural cortical stimulation for the treatment of focal epilepsy originating from eloquent cortex. *Epilepsia* **2014**, *55*, e18–e21. [CrossRef] [PubMed]

37. Lundstrom, B.N.; Worrell, G.A.; Stead, M.; VanGompel, J.J. Chronic subthreshold cortical stimulation: A therapeutic and potentially restorative therapy for focal epilepsy. *Exp. Rev. Neurother.* **2017**, *17*, 661–666. [CrossRef] [PubMed]

38. Bergey, G.K.; Morrell, M.J.; Mizrahi, E.M.; Goldman, A.; King-Stephens, D.; Nair, D.; Srinivasan, S.; Jobst, B.; Gross, R.E.; Shields, D.C.; et al. Long-term treatment with responsive brain stimulation in adults with refractory partial seizures. *Neurology* **2015**, *84*, 810–817. [CrossRef] [PubMed]

39. Schweitzer, A.; Wright, S. Effects on the knee jerk of stimulation of the central end of the vagus and of various changes in the circulation and respiration. *J. Physiol.* **1937**, *88*, 459–475. [CrossRef] [PubMed]

40. Henry, T.R. Functional imaging studies of epilepsy therapies. *Adv. Neurol.* **2000**, *83*, 305–317. [PubMed]

41. Henry, T.R. Therapeutic mechanisms of vagus nerve stimulation. *Neurology* **2002**, *59*, S3–S14. [CrossRef] [PubMed]

42. Morris, G.L.; Gloss, D.; Buchhalter, J.; Mack, K.J.; Nickels, K.; Harden, C. Evidence-based guideline update: Vagus nerve stimulation for the treatment of epilepsy: Report of the guideline development subcommittee of the American Academy of Neurology. *Neurology* **2013**, *81*, 1453–1459. [CrossRef] [PubMed]

43. Morris, G.L.; Mueller, W.M. Long-term treatment with vagus nerve stimulation in patients with refractory epilepsy. The Vagus Nerve Stimulation Study Group E01–E05. *Neurology* **1999**, *53*, 1731–1735. [CrossRef] [PubMed]

44. Elliott, R.E.; Morsi, A.; Tanweer, O.; Grobelny, B.; Geller, E.; Carlson, C.; Devinsky, O.; Doyle, W.K. Efficacy of vagus nerve stimulation over time: Review of 65 consecutive patients with treatment-resistant epilepsy treated with VNS >10 years. *Epilepsy Behav.* **2011**, *20*, 478–483. [CrossRef] [PubMed]

45. Santos, P.M. Surgical placement of the vagus nerve stimulator. *Oper. Tech. Otolaryngol.* **2004**, *15*, 201–209. [CrossRef]

46. Otsuki, T.; Kim, H.D.; Luan, G.; Inoue, Y.; Baba, H.; Oguni, H.; Hong, S.C.; Kameyama, S.; Kobayashi, K.; Hirose, S.; et al. Surgical versus medical treatment for children with epileptic encephalopathy in infancy and early childhood: Results of an international multicenter cohort study in Far-East Asia (the FACE Study). *Brain Dev.* **2016**, *38*, 449–460. [CrossRef] [PubMed]
47. Maksimow, K. Interruption of grand mal epileptic seizures by the trigeminal nerve stimulation. *Neurol. Neurochir. Pol.* **1976**, *10*, 205–208. [PubMed]
48. Soss, J.; Heck, C.; Murray, D.; Markovic, D.; Oviedo, S.; Corrale-Leyva, G.; Gordon, S.; Kealey, C.; DeGiorgio, C. A prospective long-term study of external trigeminal nerve stimulation for drug-resistant epilepsy. *Epilepsy Behav.* **2015**, *42*, 44–47. [CrossRef] [PubMed]

Article

Epilepsy: A Call for Help

Venkatraman Sadanand *

Department of Neurosurgery, Loma Linda University; Loma Linda, CA 92354, USA

Received: 15 November 2017; Accepted: 16 January 2018; Published: 28 January 2018

Abstract: Epilepsy is a considerable individual and social economic burden. In properly selected patients, epilepsy surgery can provide significant relief from disease, including remission. However, the surgical treatment of epilepsy lags in terms of knowledge and technology. The problem arises due to its slow adaptation and dissemination. This article explores this issue of a wide treatment gap and its causes. It develops a framework for a rational decision-making process that is appropriate for extant circumstances and will result in the speedy delivery of surgical care for suitable patients with medically intractable epilepsy.

Keywords: epilepsy; epilepsy surgery; decision analysis; treatment gap; game theory; economics; efficiency; resource allocation; return on investment

1. Introduction

Epilepsy is known to be a devastating disease not only for the patient but also for the patient's family and for society. If left untreated, the long-term impact can be significant. If treatment can result in being seizure-free or reduce the severity or frequency of seizures, then it may be worth treating. In fact, surgical treatment is worthwhile, but not everyone who is a good surgical candidate gets timely surgery for their epilepsy. The wait times to see a neurosurgeon are often large. In most countries, the time from diagnosis to referral to an epilepsy center ranges from 18.9 [1] to 20 years [2].

This paper explores why this might be the case. Physicians tend to rightly believe that surgical decision is based solely on medical evidence. This paper argues that this alone is not sufficient. In fact, the eventual decision for surgery of a patient is the end step in a complex decision-making process involving multiple stakeholders. This paper therefore has two tasks: first, to identify the stakeholders and their incentives; second, to model the complex interactions among this group of entities with vested interests.

For this purpose, the present paper will consider two factors that characterize this analysis. First, it postulates that the primary objective of individual stakeholders is different for each stakeholder and may conflict with those of others. From a purely public health perspective, the objective is to maximize the return on investment (ROI) while maximizing public welfare. Using the ROI approach, we will compare ROI in epilepsy surgery as the use case with ROI in glioblastoma multiforme (GBM) as a control case. The control case could be any disease with postulated lower ROI and higher investments than epilepsy surgery. Second, the complex interactions among stakeholders with diverse objectives working in an environment of strategic interactions is impossible to model using standard decision analysis wherein individual stakeholders are working with fixed outcomes of their actions. In other words, present models portray an individual stakeholder making decisions under the assumption that their decisions directly impact *only* the final outcome. In particular, they do not account for the fact that, in addition, their decisions will also have an impact upon the decisions of other stakeholders who are involved, which will alter the final outcomes. Thus, all stakeholders take actions that affect all other stakeholders, and this inevitably leads to strategic decision-making by each while being cognizant of the interdependence on others, until an equilibrium is reached where everyone obtains an outcome. This is not like previous models where a given fixed net outcome is shared by all the stakeholders (denoted sometimes as a problem of dividing a fixed pie. Those problems are referred to as zero-sum games. The current problem we analyze is not a zero-sum game). In our analysis, when equilibrium is reached, each stakeholder obtains an outcome that is of a specific value to that stakeholder. What we require is a way to model the strategic interactions among multiple stakeholders when each one has an individualized objective that may conflict with those of others in the same environment. Such models of strategic decision-making were first postulated mathematically by John Nash [3] and subsequently developed further through applications to other fields [4]. This field of mathematics is called non-cooperative game theory.

This is the first time non-cooperative game theory is being used in the literature on medical decision analysis. Why is non-cooperative game theory a better tool for medical decision analysis than standard Markov decision trees? The reason is that it is more realistic and hence a more accurate model of stakeholder interaction. The stakeholders are not passive observers in the system as in the current modeling techniques in medical decision analysis. In fact, every stakeholder is a maximizer of his own objective function and, in that process, recognizes and adjusts dynamically to the behavior calculus of every other stakeholder. Each stakeholder will try to predict the behaviors of all other stakeholders in an environment of uncertainty and incomplete information about others. This is the essence of non-cooperative game theory. Economic modeling, market analysis, individual behavior analysis, group dynamics, and military decision-making have all abandoned the previous naïve models of dynamic programming with fixed incentive for players. Instead these models have evolved to game theory. This shift from fixed incentives-based interactions to a dynamic interaction of multiple players with diverse objectives who think through others' thought processes and adjust on the go is the very basis of a complex mathematical analysis leading to the Nobel Prize in Economics in 1994 [3–5]. The mathematics behind such analysis has been referred to as game theory. Even individual biological cell behavior is now seen to follow such models, that have been referred to as biological games. Thus, this paper deviates from existing literature and advocates the use of game theory for medical decision-making and considers the techniques of static and dynamic programming with fixed incentives as a source of misleading results and not a model of real behavior.

Consider the example of a low returns (ROI)–high investment disease such as glioblastoma multiforme (GBM), an aggressively malignant brain tumor. This is also a devastating disease with considerable negative impact on the individual and his or her family. Despite the best treatment options, median survival is about 18 months [6], and most patients who are diagnosed with this disease have low productive horizons ahead of them.

Current data appears to establish that, if we compare epilepsy with GBM, the incidence of GBM is lower, prevalence is lower, direct treatment costs are higher, indirect treatment costs are higher, benefits are lower, contribution to GDP is lower, and ROI is lower—yet we invest more readily in GBM treatment than in epilepsy surgery. A patient with GBM reaches the surgeon usually within a few days after diagnosis. A patient with epilepsy takes 18–20 years [1,2] from diagnosis to reach the surgeon in most countries. This is a problem.

This chapter addresses this issue, quantifies it, and suggests an approach based on game theoretic analysis to develop a methodology to take a deeper look at the nature of this treatment gap and socio-economic inefficiency.

In public health, as in public policy, economic efficiency is measured by a concept borrowed from applied mathematics called Pareto optimum [7]. When the welfare of stakeholders can be improved from a given situation without making anyone less well-off than the status quo, it is called a Pareto improvement. It is well established in the economics and public policy literature that economic efficiency requires Pareto optimal outcomes. This well-known result led to the Nobel Prize in 1972 for Kenneth Arrow [8]. More recently, it has been advocated that Pareto optimization cannot be ignored in medical decision-making and public health [9]. Game theory will model realistic interaction among stakeholders, and the theorem of Pareto optimality will yield economic efficiency. Thus, amid paucity, the creation of epilepsy centers of excellence can be Pareto optimal.

2. Problem Definition

The underlying problem explored in this chapter is best described by Table 1, where the two diseases are compared: glioblastoma multiforme (GBM) and surgical epilepsy (SE). SE is 3000 times more prevalent than GBM. There are 400 times more patients who are candidates for SE than for GBM. Total expenditures on research and treatment is roughly the same for both diseases. However, the benefits of surgery per year for a patient with SE is 80,000 times that of a patient with GBM.

In Table 1, the direct costs for epilepsy include neurology specialist consultations, primary care visits for reasons related to the disease, number and type of diagnostic tests performed (basic blood test, analysis of anti-epileptic drug levels, brain Computed Tomography (CT) scan, brain Magnetic Resonance Imagine (MRI), brain Single Photon Emission Computed Tomography (SPECT) and Positron Emission Tomography (PET), Electrocardiogram (ECG), Electroencephalogram (EEG), EEG with sleep deprivation, Holter EEG, Video Electroencephalogram (VEEG), and Holter ECG), days of hospitalization for diagnostic work up, and the treatment administered. Non-medical direct costs include the use of transport to and from hospital and psycho-educational and social support. Costs of anticonvulsant medications were the highest factor in direct costs. Furthermore, the indirect costs of epilepsy are higher than direct costs in most studies. This is because of the value of a productive person in society even if only accounting for a small probability of diminished abilities after surgery.

In the case of GBM, direct costs include consultations with primary care physicians, neurology or neurosurgery specialists, and diagnostic tests such as CT scans, brain MRI, as well as ECGs and blood tests. Once the diagnosis of a brain tumor is made, the patient goes immediately (during the same admission in most cases) to surgery. Surgery and recovery then become direct costs. The last component of direct costs arise from post-operative care, chemotherapy, radiation therapy, rehabilitation, and a repeat of direct costs for each recurrence that is treated. Indirect costs for GBM are lower than for epilepsy because of the limited life expectancy post-diagnosis. In addition, the average age of diagnosis is greater in GBM compared to epilepsy. Hence the span of productive life lost is also less than for epilepsy.

Table 1. Cost-Benefit Comparison of GBM versus SE.

Disease	Prevalence (Global)	Incidence (Global)	Surgical Candidates (Global)	Median Survival	Age Distribution of Disease	Costs: Direct/Year in US$ (Research, Prevention, Treatment)	Costs: Indirect/Year in US$ (Individual, Family, Society)	Benefits: Per Year in US$ (Estimated Value of Productive Life)	Benefit/Cost Ratio (ROI Equivalent)
Glioblastoma Multiforme (GBM)	1.6×10^4 [10]	5.0×10^4 [11]	5.0×10^4 [11]	1.5 years [6]	45–70 [6]	1045×10^8 [12,13]	390×10^8 [14]	20×10^8 [15]	0.01
Surgical Epilepsy (SE)	50×10^6 [16]	2.4×10^6 [16]	20×10^6 [16]	60 years [17]	Multimodal [18]	250×10^8 [19]	700×10^8 [19,20]	160×10^{12} [15]	1684

Waiting times for GBM patients, interestingly, are extremely low compared to epilepsy patients. Median time from diagnosis to surgery is less than 1 week and median times from surgery to adjuvant therapy is 27 days. This is despite the costs reported [13] to be between $50,000 and $92,000 per Quality Adjusted Life Years (QALY), barely making it over the accepted cost-effectiveness thresholds of $50,000 per QALY.

In the United States of America, 369 resective epilepsy surgeries are being carried out per year on the average, whereas there are about 1 million surgical candidates for epilepsy surgery [21]. In India, with a population approximately 4 times that of the United States, only 734 resective epilepsy surgeries are carried out per year when there are 2.5 million surgical candidates [22]. In China, there are about 1.8 million surgical candidates for epilepsy surgery and only about 2500 epilepsy surgeries are carried out per year [23]. Based on the data from these studies [21,23], China has roughly twice the number of surgical candidates but performs eight times more epilepsy surgeries per year. However, one must be cautious before this finding is applauded. Both countries are still woefully short of treating the number of epilepsy surgery candidates. There is a considerable treatment gap impacting individual and social welfare and that treatment gap is the subject of this paper. It is even more surprising that there is such a treatment gap when the seizure-free rate averaging over all types of seizure surgeries is 68% and about 90% for certain kind of seizure foci [24].

We have thus far examined the treatment gap, the cost–benefit ratio, and the time taken from diagnosis to surgical attention as measures of this seeming social inefficiency. We may also consider other measures such as the cost-effectiveness ratio and the return on investment (ROI).

Any medical intervention that extends lives is measured by the cost-effectiveness ratio. It is the ratio of costs divided by number of years gained. This does not account for the value ascribed to those years. The smaller the number, the more worthwhile the intervention. To calculate this, it should be noted that investment for GBM research and treatment extends life by about 1.5 years [6], while investment for epilepsy extends good quality of life by about 40 years on average. This is a conservative estimate due to the multimodal distribution of incidence (computed from [13] by taking the scaled mean of histograms and subtracting that from the WHO 2015 estimate of life expectancy).

Cost-Effectiveness Ratio (CER) = Direct and Indirect Costs in US$/Benefits in Years Gained

For GBM, the CER: $1435 \times 10^8/1.5 = 956 \times 10^8$.

For Epilepsy, the CER is: $1700 \times 10^8/40 = 42.5 \times 10^8$.

The Benefit Cost Ratio (BCR) is like the CER but considers the value of life lived. This then becomes a proxy for the ROI and has been computed and is shown in the last column of Table 1.

While there are several other measures one could use, such as the QALY, the point is clear: SE research and treatment, compared to GBM research and treatment, is more efficient, more efficacious, better for the patient, and better for the society. However, SE treatment is gravely underutilized. Thus, how is this market inefficiency explained and how can this be addressed? A purely medical decision-making is an evidence-based decision. On that basis alone, there is no paucity of evidence to show that, for the properly identified candidate, epilepsy surgery is the right option rather than medical management. However, this would ignore the reality of a medical economy. One must review the incentives and disincentives of all stakeholders in this game of budget allocation for surgery to advance the patient from diagnosis to surgery. Just because medical evidence shows that epilepsy surgery is better for properly selected patients compared to medical management, it does not mean that those patients will receive surgery in an environment where there are several stakeholders with conflicting and diverse goals.

3. The Stakeholders, Their Incentives and Pitfalls

Surgical decision-making in epilepsy involves several stakeholders along the path from diagnosis to surgery. The first is the physician. The second is the patient and his or her family. The third is hospital administration. The fourth is the insurance industry. The fifth is society.

The physician's decisions are typically evidence-based. This is a necessary condition for surgery but not sufficient. However, their role can push to decrease the large gap of almost 20 years from diagnosis to surgery and, once cleared for surgery, to push for appropriate budget allocations to enable surgery. The doctor, in most epilepsy centers, must take on the role of a physician and that of a patient advocate.

Next, the patient and his family are dealing with a disease that not only results in a loss of income and productivity but also the social stigma and discrimination associated with epilepsy. Therefore, they need to be educated in the social and quality of life advantages of epilepsy surgery. Such patient education is glaringly lacking until the patient somehow makes his first contact with a doctor. In some countries, epileptologists are usually based in urban centers while many epilepsy patients are in rural areas.

Hospital administrations must balance budgets and face multiple demands from different subspecialties. Resource allocation then becomes an issue of educating administrators about the cost-effectiveness and efficacies inherent in epilepsy surgery, even when compared to other specialties that may have their own demands for fixed budgets, finding a proper fit with corporate missions and medicolegal experiences.

Insurance industry needs to be made aware of the cost–benefit analysis of surgery versus the medical care of epilepsy patients. In computing these costs and benefits, care must be taken to account for the discounted present values along with the non-zero rates of return. Insurance companies therefore must now weigh the cost of surgical care against the discounted present value of the future cost of medical care. Social welfare and lost productivity are often not in their optimization problem and may be considered externalities [25]. Externalities impose costs on other stakeholders but have no impact on the given stakeholder's mathematical profit maximization function.

The last stakeholder is society. How does one account for the value of epilepsy surgery to society? Epilepsy surgery creates productive people who can live healthy lives. That is the very basis of measures such as BCR and ROI. Thus, the impact of epilepsy surgery on this last stakeholder is measured by the value of an individual on society. The current belief is that the value of an individual's life is measured by the market value of the goods and services produced by that individual. However, this may be incorrect. The value of an individual's productivity to society may exceed the actual productive output of the individual due to social multiplier effects and what is known as Okun's law [26,27], and this must also be accounted for in computing society's objectives in this interactive strategic decision analysis.

4. Decision Analysis

What we have now is a complex problem of resource allocation with the following characteristics: the existence of multiple strategic stakeholders (so current models of a single stakeholder such as a physician, making medical evidence-based decisions alone is an inadequate model), diverse goals (so a multiple stakeholder decision analysis wherein all stakeholders have the same goal of patient welfare is also inadequate), computational complexities of payoffs to each stakeholder (so a model of multiple stakeholders with fixed objectives is inadequate), the interdependence of multiple stakeholders wherein one stakeholder's decisions affects not only his own but also the outcomes of others (hence a model where a stakeholder's choices only affect his outcome is inadequate) and the freedom to choose what is best for each stakeholder (so a model of decision analysis that ignores the strategic freedom to choose for each stakeholder is also inadequate). In this environment, it would not be helpful to model this situation as a simple static evidence-based decision for the surgeon. Firstly, the surgeon is not the only stakeholder. Secondly, not every stakeholder has the same objective and incentives as the surgeon. Lastly, not every stakeholder fully knows what others know. This is the reason why models of what ought to be the outcome (surgery, in this case) often differs from what becomes the outcome (treatment gap, in this case).

Brain Sci. **2018**, *8*, 22

The process of medical decision-making is therefore best described and computed by what is known as strategic game theory [28]. Each stakeholder is working in an environment of incomplete information (not everyone knows what others fully know) about the incentives faced and considered by other stakeholders and uncertainty about their actual decisions. A decision analysis will create a game of imperfect information in which all the stakeholders choose strategies to maximize their gains under uncertainty in the environment and under uncertainty about others' choices. Such complex interactions may involve coalition forming among the stakeholders and may even result in sharing gains from the outcome.

In such games, every player first defines what his or her goal is and what would be the mathematical representation of those goals. It then seeks to maximize that mathematical entity while accounting for other stakeholders' similar thought processes. Outcomes in such situations are called equilibria of games [29]. Since every player functions under some degree of uncertainty about the other players' motives and moves, the final action of each stakeholder is dictated by what is known as Bayesian expected utility and interactive equilibrium [29]. This is the decision-making process used by most Forbes 100 businesses, and it may be about time that medical decision-making follows this approach by considering the strategic nature, the informational asymmetries, uncertainties in the environment and the diverse goals of stakeholders.

5. Conclusions

There is medical evidence that, for carefully selected candidates, surgery is better than medical management for patients with medically intractable epilepsy. However, SE is underutilized and there is a treatment gap. SE patients must wait a long time after diagnosis to be referred to a neurosurgeon, resulting in patients who do not receive the surgeries they need. This is not the case for GBM. Patients with diagnosed GBM have waiting times for referral for surgery of a few days, whereas epilepsy patients can wait years. In addition, almost every patient with a first-time diagnosis of possible GBM undergoes surgery.

The problem may stem from complex interactions of several stakeholders. This paper advocates a method of analysis, called mathematical game theory, that has been used in the fields of business, economics, public policy, political science, and engineering and biology.

Decision-making for GBM surgery is much easier as all stakeholders appear to focus on the short life expectancy after the diagnosis of GBM and the apprehensive determination to do something to change that.

It would seem, that fear of death drives decisions differently than hope of improving the quality of long-term life.

Acknowledgments: The author has received no funding for this research and its publication.

Conflicts of Interest: The author declares no conflict of interest.

References

1. Martínez-Juárez, I.E.; Funes, B.; Moreno-Castellanos, J.C.; Bribiesca-Contreras, E.; Martínez-Bustos, V.; Zertuche-Ortuño, L.; Hernández-Vanegas, L.E.; Ronquillo, L.H.; Rizvi, S.; Adam, W.; et al. A comparison of waiting times for assessment and epilepsy surgery between a Canadian and a Mexican referral center. *Epilepsia Open* **2017**, *2*, 453–458. [CrossRef]
2. Thornton, J.G.; Lilford, R.J.; Johnson, N. Decision analysis in medicine. *Br. Med. J.* **1992**, *304*, 1099. [CrossRef]
3. Nash, J.F., Jr. Equilibrium points in n-person games. *Proc. Natl. Acad. Sci. USA* **1950**, *36*, 48–49. [CrossRef] [PubMed]
4. Basar, T.; Olsder, G.I. *Dynamic Noncooperative Game Theory*, 2nd ed.; Society for Industrial and Applied Mathematics (Classics in Applied Mathematics): Philadelphia, PA, USA, 1999.
5. Reinhard Selten—Facts. *Nobelprize.org*; Nobel Media. Available online: http://www.nobelprize.org/nobel_prizes/economic-sciences/laureates/1994/selten-facts.html (accessed on 12 December 2017).

6.   Johnson, D.R.; O'Neil, B.O. Glioblastoma survival in the United States before and during the temozolomide era. *J. Neurooncol.* **2012**, *107*, 359–364. [CrossRef] [PubMed]
7.   Saule, C.; Giegerich, R. Pareto optimization in algebraic dynamic programming. *Algorithms Mol. Biol.* **2015**, *10*, 22. [CrossRef] [PubMed]
8.   Kenneth, J. Arrow—Facts. *Nobelprize.org*; Nobel Media. Available online: http://www.nobelprize.org/nobel_prizes/economic-sciences/laureates/1972/arrow-facts.html (accessed on 12 December 2017).
9.   Dewar, D. *Essentials of Health Economics*, 2nd ed.; Jones & Bartlett Learning: Burlington, MA, USA, 2017.
10.   Glioblastoma. Available online: http://www.orpha.net/consor/cgi-bin/OC_Exp.php?Expert=360 (accessed on 12 December 2017).
11.   Louis, D.N. *WHO Classification of Tumours of the Central Nervous System*; International Agency for Research on Cancer: Lyon, France, 2007.
12.   Wasserfallen, J.-B.; Ostermann, S.; Leyvraz, S.; Stupp, R. Cost of temozolomide therapy and global care for recurrent malignant gliomas followed until death. *Neuro Oncol.* **2005**, *7*, 189–195. [CrossRef] [PubMed]
13.   Raizer, J.J.; Fitzner, K.A.; Jacobs, D.I.; Bennett, C.L.; Liebling, D.B.; Luu, T.H.; Trifilio, S.M.; Grimm, S.A.; Fisher, M.J.; Haleem, M.S.; et al. Economics of malignant gliomas: A critical review. *J. Oncol. Pract.* **2014**, *11*, e59–e65. [CrossRef] [PubMed]
14.   Blomqvist, P.; Lycke, J.; Strang, P.; Törnqvist, H.; Ekbom, A. Brain tumours in Sweden 1996: Care and costs. *J. Neurol. Neurosurg. Psychiatry* **2000**, *69*, 792–798. [CrossRef] [PubMed]
15.   How Much Will We Pay for a Year of Life? Available online: https://www.gsb.stanford.edu/insights/how-much-will-we-pay-year-life (accessed on 12 December 2017).
16.   World Health Organization. *Atlas: Epilepsy Care in the World*; World Health Organization Press: Geneva, Switzerland, 2005.
17.   Gaitatzis, A.; Johnson, A.L.; Chadwick, D.W.; Shorvon, S.D.; Sander, J.W. Life expectancy in people with newly diagnosed epilepsy. *Brain* **2004**, *127*, 2427–2432. [CrossRef] [PubMed]
18.   Kotsopoulos, I.A.; Van Merode, T.; Kessels, F.G.; de Krom, M.C.; Knottnerus, J.A. Systematic Review and Meta-analysis of Incidence Studies of Epilepsy and Unprovoked Seizures. *Epilepsia* **2002**, *43*, 1402–1409. [CrossRef] [PubMed]
19.   Pato-Pato, A. Evaluation of the Direct Costs of Epilepsy. *J. Neurol. Neurosci.* **2013**, *4*, 3. [CrossRef]
20.   Kotsopoulos, I.A.W.; Evers, S.M.A.A.; Ament, A.J.H.A.; De Krom, M.C.T.F.M. Estimating the Costs of Epilepsy: An International Comparison of Epilepsy Cost Studies. *Epilepsia* **2001**, *42*, 634–640. [CrossRef] [PubMed]
21.   Englot, D.J.; Ouyang, D.; Garcia, P.A.; Barbaro, N.M.; Chang, E.F. Epilepsy surgery trends in the United States, 1990–2008. *Neurology* **2012**, *78*, 1200–1206. [CrossRef] [PubMed]
22.   Rathore, C.; Radhakrishnan, K. Epidemiology of Epilepsy Surgery in India. *Neurol. India* **2017**, *65*, 52–59. [CrossRef]
23.   Xu, L.; Xu, M. Epilepsy Surgery in China: Past, Present and Future. *Eur. J. Neurol.* **2009**, *17*, 189–193. [CrossRef] [PubMed]
24.   Spencer, S.S.; Berg, A.T.; Vickrey, B.G.; Sperling, M.R.; Bazil, C.W.; Shinnar, S.; Langfitt, J.T.; Walczak, T.S.; Pacia, S.V. Predicting Long-Term Seizure Outcome after Resective Epilepsy Surgery: The Multicenter Study. *Neurology* **2005**, *65*, 912–918. [CrossRef] [PubMed]
25.   Arrow, K. The organization of economic activity: Issues pertinent to the choice of market versus non-market allocation. In *Public Expenditure and Policy Analysis*; Haveman, R.H., Margolis, J., Eds.; United States Congress Joint Economic Committee: Washington, DC, USA, 1970; pp. 59–73.
26.   Okun, A.M. *Potential GNP, Its Measurement and Significance*; Cowles Foundation, Yale University: New Haven, CT, USA, 1962.
27.   Gordon, R.J. *Productivity, Growth, Inflation and Unemployment*; Cambridge University Press: Cambridge, England, 2004.
28.   Von Neumann, J.; Morgenstern, O. *Theory of Games and Economic Behavior*; Princeton University Press: Princeton, NJ, USA, 1944.
29.   Harsanyi, J.C.; Selten, R. *A General Theory of Equilibrium Selection in Games*; MIT-Press: Cambridge, MA, USA, 1988.

brain sciences

MDPI

Review

Surgical Considerations of Intractable Mesial Temporal Lobe Epilepsy

Warren W. Boling

11234 Anderson Street, Room 2562-B, Loma Linda, CA 92354, USA; wboling@llu.edu; Tel.: +1-909-558-4479

Received: 1 February 2018; Accepted: 15 February 2018; Published: 20 February 2018

Abstract: Surgery of temporal lobe epilepsy is the best opportunity for seizure freedom in medically intractable patients. The surgical approach has evolved to recognize the paramount importance of the mesial temporal structures in the majority of patients with temporal lobe epilepsy who have a seizure origin in the mesial temporal structures. For those individuals with medically intractable mesial temporal lobe epilepsy, a selective amygdalohippocampectomy surgery can be done that provides an excellent opportunity for seizure freedom and limits the resection to temporal lobe structures primarily involved in seizure genesis.

Keywords: temporal lobe epilepsy; selective amygdalohippocampectomy; epilepsy surgery; mesial temporal lobe epilepsy

1. Introduction

Temporal lobe epilepsy (TLE) affects a substantial number of individuals with medically intractable epilepsy. TLE is the most common operated epilepsy, and the majority of patients with localization related epilepsy seen in tertiary epilepsy centers have TLE [1,2]. Although TLE is not the most common epilepsy. TLE in the general population of Minnesota has been estimated at about 10.4 per 100,000 this is compared with 54.3 per 1000 incidence of epilepsy in the population as a whole [3].

The frequent occurrence of intractable epilepsy in the temporal lobe bears witness to the highly epileptogenic nature of the limbic structures that comprise the mesial portion of the temporal lobe. The mesial part of the temporal lobe is richly connected with surrounding extra-temporal cortical regions especially the orbitomesial frontal lobe via the uncinate fasciculus and the fornix carries fibers from the hippocampus that project to the anteromesial frontal lobe and anterior nucleus of the thalamus [4,5]. The mesial temporal structures are highly connected as well with the anterolateral neocortical temporal lobe. Therefore, due to the strong connections of the mesial temporal structures with the anterior and lateral temporal lobe in addition to other limbic regions, TLE most commonly manifests the semiology of staring and automatisms regardless of the seizure onset zone in lateral or mesial structures of the temporal lobe.

Despite the frequent occurrence and intractable nature of temporal lobe epilepsy, it responds very well to surgery with high rates of resulting seizure freedom and the risks of surgery are quite low [6–8]. The decision to consider a patient with medically intractable TLE for surgery has become much clearer nowadays since the International League Against Epilepsy (ILAE) defined medical intractability as the failure of adequate trials of two tolerated, appropriately chosen and used antiepileptic drugs [9]. A clear definition of intractable epilepsy combined with the excellent results of surgery for symptomatic epilepsy especially arising in the temporal lobe, has resulted in numerous consensus reports affirming that individuals with drug resistant epilepsy should be evaluated in a comprehensive epilepsy program to identify opportunities for surgical cure [10].

Temporal lobe epilepsy can be categorized into one of two types based on the anatomical region of seizure onset. Seizures that originate from the temporal cortex lateral to the collateral sulcus are defined as lateral or neocortical epilepsy, and seizures that have a focus of onset medial to the collateral sulcus are named mesial temporal lobe epilepsy (MTLE). The International League Against Epilepsy recognizes there are sufficient distinguishing characteristics for MTLE to be considered a distinct syndromic entity [11]. Although the two categories of TLE frequently share the same limbic semiology [11–13], in general, MTLE more commonly displays the epigastric, cephalic or experiential aura, loss or awareness, staring, automatisms and posturing that are typical temporal lobe seizure patterns and are the result of a seizure prominently involving the limbic structures [14]. The imaging finding that defines MTLE is atrophy and sclerosis of the hippocampus, so-called mesial temporal sclerosis (MTS) [15]. The childhood history of patients with MTLE commonly includes the presence of childhood febrile convulsions, especially of a prolonged and complicated nature [16]. Individuals with neocortical epilepsy are more likely to manifest signs related to peri-Sylvian structures such as a simple auditory hallucinations or, in the dominant hemisphere, postictal aphasia [17,18]. Although neocortical epilepsy often spreads along fibers richly connected with the mesial temporal structures manifesting the signs and symptoms of limbic involvement, which are the semiological features of MTLE as well.

Due to the considerable overlap in semiology between the two categories of temporal lobe epilepsy, multiple noninvasive data elements must converge to localize the TLE focus to the mesial or lateral structures of the temporal lobe [19,20]. However, when patient history, seizure semiology, EEG localization, and imaging findings point to MTLE, there is a high degree of confidence in the diagnosis [21,22]. In a minority of patients, intracranial electrode monitoring may be required to investigate lateralization of the seizure onset to a temporal lobe [23] or to confirm temporal lobar localization in one hemisphere [24].

2. Surgery of Temporal Lobe Epilepsy

2.1. Temporal Lobectomy

Temporal lobectomy is a frequently a misused term that is defined as removal of the entire temporal lobe including the mesial structures. A complete removal of the temporal lobe is rarely performed today. The description of lobectomy to describe a resection should be reserved for the unique situations in which an entire anatomical temporal lobe actually is removed. The standard resections done for temporal lobe epilepsy are preferably described as an anterior temporal resection or cortico-amygdalohippocampectomy (CAH), which more accurately describes the anatomic structures that are removed.

2.2. Cortico-Amygdalohippocampectomy (CAH)

This is the standard temporal resection commonly performed at most epilepsy centers. It essentially corresponds to an anterior temporal neocortical resection followed by a removal of the temporal mesial structures. Variations on this approach have been described including anatomically standardized resections [24–26] as well as a tailored type of operation [27].

The anatomical resection approach consists, in the dominant hemisphere of the resection along the superior temporal gyrus (T1) posteriorly to the level of the pre-central sulcus, which corresponds to about 3.5–4 cm from the temporal pole. In addition, in the non-dominant hemisphere the T1 removal is delineated posteriorly by the central sulcus, which is about 4–4.5 cm from the temporal pole. Figure 1 The cortex and white matter lateral to the collateral sulcus is removed en bloc as a specimen in the first step. Next, the temporal horn is opened in order to identify the mesial temporal structures, which are disconnected and removed as described below for the selective amygdalohippocampectomy (SAH).

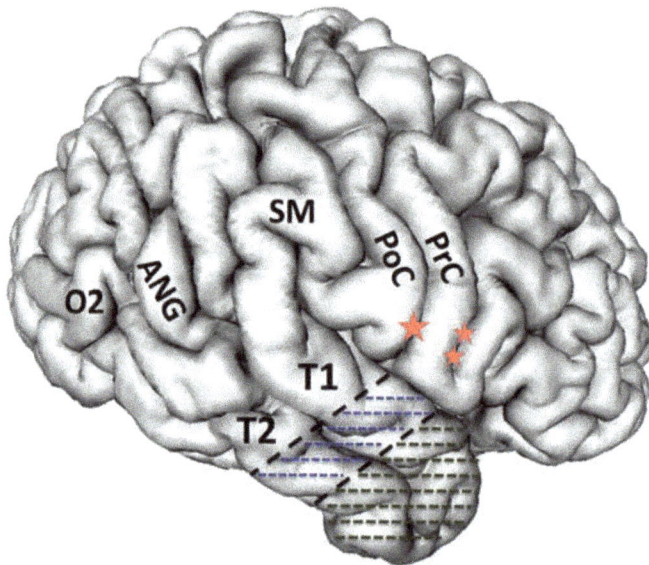

Figure 1. 3-D reconstruction of the non-dominant right hemisphere. The dashed line illustrates the anatomical resection approach of cortico-amygdalohippocampectomy. In the non-dominant hemisphere, the resection is taken posteriorly to the level of the central sulcus along T1. In the dominant hemisphere, the T1 resection is no further posterior than the precentral sulcus in order to respect and preserve posterior language areas. Single star = central sulcus, double star = pre-central sulcus, T1 = superior temporal gyrus, T2 = middle temporal gyrus, PoC = post central gyrus, PrC = precentral gyrus, SM = supramarginal gyrus, ANG = angular gyrus, O2 = second occipital gyrus, which is the gyral continuum of T2 in the occipital lobe.

2.3. Cortico-Amygdalectomy (CA)

As the name implies, CA consists of an anterior temporal resection along with the amygdala and uncus. The remaining hippocampal formation, parahippocampus, and entorhinal cortex are spared. The lateral neocortical resection is limited posteriorly as described above for CAH. Figure 1 In the CA, the cortical resection is limited medially by the collateral sulcus thus preserving parahippocampus and entorhinal cortex. The anterior temporal horn is opened for the purpose of identifying intraventricular landmarks to confirm that the hippocampus and the adjacent parahippocampal gyrus are left intact. Opening the ventricle also facilitates removal of the amygdala and subpial emptying of the uncus. This surgical approach is most useful in patients who are at risk for a functional memory decline from removal of the hippocampal complex or have failed the ipsilateral intracarotid amobarbital test for memory. Kim, et al. identified with neuropsychological testing before and after surgery that this surgical approach avoids memory decline, yet provides a good opportunity for seizure freedom in patients with TLE when the hippocampus retains significant memory function [28,29].

2.4. Selective Amygdalohippocampectomy (SAH)

The SAH surgical approach is based on the understanding that the seizure focus of MTLE is confined to the mesial temporal structures. Hughlings Jackson first illustrated this concept when he described a patient with a lesion in the uncus causing psychomotor seizures (Jackson called a dreamy state), which was the first description of the role the temporal lobe structures in human epilepsy [30]. More modern experimental studies have subsequently identified the critical role of the mesial temporal structures in experimental and human TLE [31–36]. Additionally, seizure freedom results realized from many patients operated for MTLE with SAH confirms the importance of the mesial temporal and limbic structures in MTLE [35,37–40]. Because MTLE represents a sufficiently distinct pathophysiological entity, the International League Against Epilepsy has concluded that it does represent a specific epilepsy syndrome [11].

The SAH surgical approach was first described by the Brazilian neurosurgeon, Paolo Niemeyer in the 1950's to treat TLE arising from the mesial temporal structures. Niemeyer described the innovative surgical technique he called "transventricular amygdala-hippocampectomy" and presented the rational for a selective approach to surgery for temporal lobe epilepsy at the International Colloquium on Temporal Lobe Epilepsy in 1957 that took place at the National Institutes of Health [41]. Figure 2 Subsequently, the importance of MTLE in medically intractable TLE has been established by reports confirming the utility of SAH for the treatment of MTLE, [38,42–44] and MR imaging now greatly enables the diagnosis of patients with MTLE by defining the anatomical correlate, which is hippocampal sclerosis [11,15].

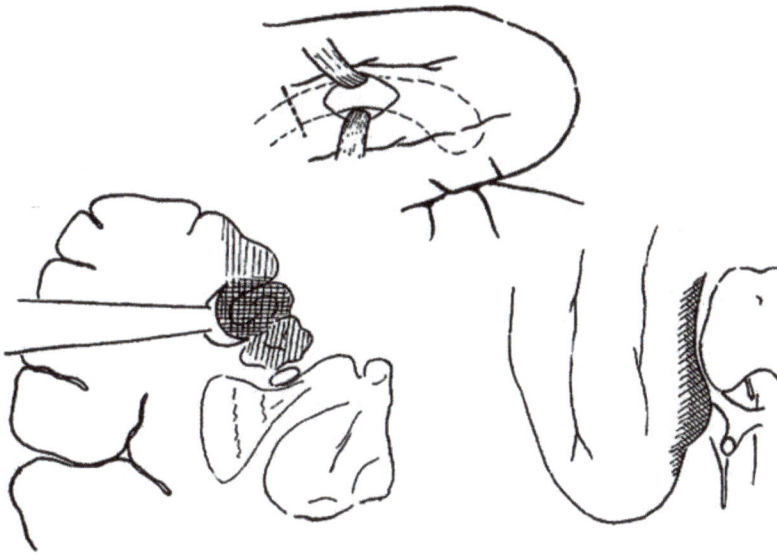

Figure 2. Reproduced with permission from Temporal Lobe Epilepsy: A Colloquium Sponsored by the National Institute of Neurological Diseases and Blindness, National Institutes of Health, Bethesda, Maryland, in Cooperation with the International League Against Epilepsy [45] Copyright©1959 American Medical Association. All rights reserved. In his presentation at the colloquium, Niemeyer described the convincing evidence from experimental research and clinical experience that the mesial temporal structures have a critical role in epileptogeneis of TLE. Niemeyer described a surgical technique he developed to selectively remove the mesial temporal structures via a trans-T2 trans-ventricular approach.

Niemeyer's description of a white matter corridor through the middle temporal gyrus (T2) to the temporal horn to selectively remove the mesial temporal structures was a very different approach than the anterior temporal resection popular at the time. In fact, research carried out in Montreal by Scoville and Milner [46] and Penfield and Milner [47], which had stressed the important role of the hippocampus in memory, as well as the earlier described Klüver-Bucy syndrome had influenced centers surgically treating epilepsy in North America to preserve the hippocampus. Concerns about memory were reflected in the first publication of a series of patients operated for TLE in which Penfield and Flanigin described hippocampus removal in only 2 of 32 patients operated [48]. However, soon after the Penfield and Flanigin publication, experimental evidence became overwhelming that the mesial temporal structures had an important role in epileptogenesis of TLE [31,32,34,49,50]. Penfield and Jasper demonstrated a few years after the Penfield and Flanigin publication that further removal of the hippocampus could convert a failed TLE surgery into a success [36]. In Britain, Murray Falconer was an early advocate of including the hippocampus in the resection for TLE. He spoke frequently about removal of the mesial temporal structures to provide the best opportunity for seizure freedom in TLE, and Falconer identified the importance of mesial temporal sclerosis (he coined the term) in the pathogenesis of TLE [51]. William Feindel in Montreal demonstrated an important role of the amygdala in epileptogenesis and in generation of the typical automatism semiology of TLE [52].

Yasargil, et al. generated considerable interest in SAH for MTLE after the description of a selective surgical approach using the trans-Sylvian fissure route to remove the mesial temporal structures [53]. In the technique described by Yasargil, the Sylvian fissure is opened to expose the inferior aspect of the circular sulcus of the insula. An incision in the circular sulcus between middle cerebral temporal opercular arteries and dissection through the white matter of the temporal stem will open the temporal ventricular horn providing sufficient exposure to identify then remove the hippocampal formation and amygdala. Figure 3A,B An advantage of this surgical approach to SAH is to minimize white matter dissection required to access the mesial temporal structures as compared with Niemeyer's technique. The disadvantages of the trans-Sylvian approach are mostly related to the demanding technical requirements of widely opening the Sylvian fissure and working between the middle cerebral vessels within the fissure. Manipulation of the middle cerebral arteries during surgery in addition to the extra-pial resection of the hippocampus are technique related aspects of the procedure that may increase the risk of vascular injury [54].

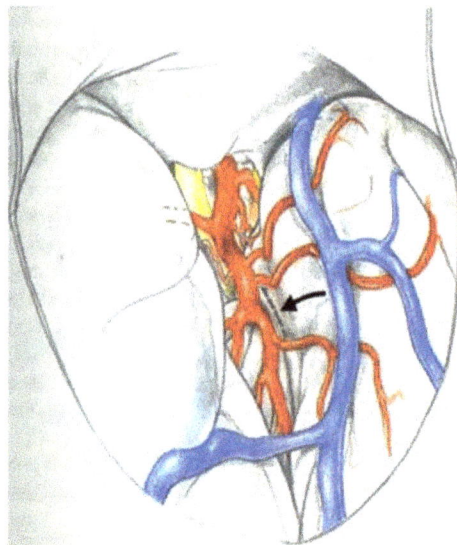

(A)

Figure 3. *Cont.*

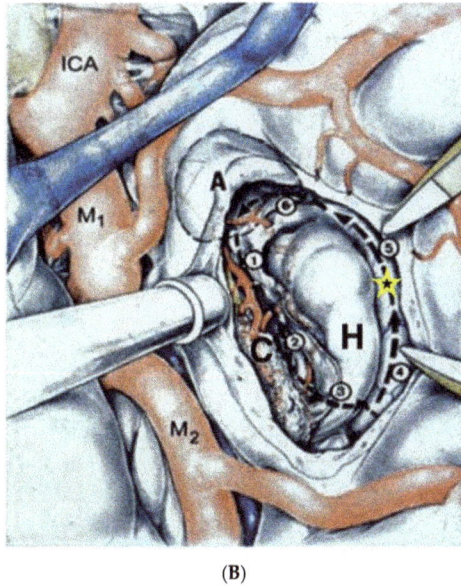

(**B**)

Figure 3. Reprinted and Modified by permission from Springer Nature: Adv Tech Stand Neurosurg, Yaşargil MG, Teddy PJ, Roth P. Selective amygdalohippocampectomy: operative anatomy and surgical technique. Copyright, 1985 [55]. (**A**) The Sylvian fissure is opened widely to expose the inferior delineation of the insula called the circular sulcus with the planned white matter incision marked in a dashed line between two middle cerebral opercular temporal arteries; (**B**) Surgeon's view into the opened ventricle illustrating the choroid plexus (C), hippocampus (H), collateral eminence (star), middle cerebral vessels (M1 and M2), and amygdala (A). Numbers represent steps in the extra-pial removal of mesial temporal structures: coagulation and division of the hippocampal vessels (steps 1 and 2), disconnection of the hippocampal body from the hippocampal tail (step 3), opening the collateral eminence to empty and remove the parahippocampus (steps 4 and 5), and remove uncus and amygdala (step 6).

Hori contributed importantly to the SAH literature by describing a subtemporal amygdalohippocampectomy technique to perform SAH [56]. This approach includes at least partial removal of fusiform gyrus in order to gain access to the parahippocampal gyrus and mesial structures via a subtemporal route. Figure 4 Hori's original description was extra-pial resection of the mesial temporal structures but later he modified the approach to a subpial resection technique. To minimize retraction on the temporal lobe, Hori drilled down the roof of the external auditory meatus. A preoccupation with the subtemporal approach is retraction injury of the anastomotic vein of Labbé, which Hori was able to successfully preserve using techniques described in the skull base surgery literature [57]. The advantages of the subtemporal approach that Hori described were preservation of visual fields, he identified no superior quadrantanposia after surgery, and the temporal lobe is not disconnected in contradistinction to Yasargil's trans-Sylvian approach that requires dissecting through the temporal stem to access the mesial temporal structures. The disadvantages of the subtemporal approach are related mostly to the extended resection of the fusiform gyrus that is not required in other SAH approaches as well as an inherent risk of injury to the anastomotic draining vein of Labbé from elevating the temporal lobe with a retractor, which is a particular concern in the dominant temporal lobe.

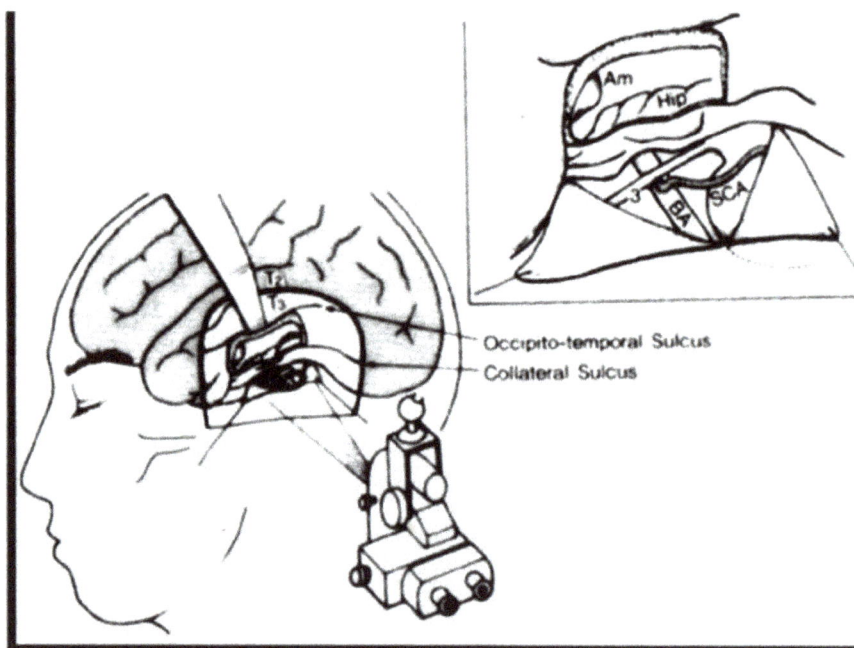

Figure 4. Reproduced with permission from: Figure 1. Tomokatsu Hori et al. Subtemporal Amygdalohippocampectomy for Treating Medically Intractable Temporal Lobe Epilepsy. *Neurosurgery* (1993) 33 (1): 50–57, [56]. Published by Oxford University Press on behalf of the Congress of Neurological Surgeons. Hori developed the subtemporal approach to SAH. The illustration shows a retractor elevating T3 (inferior temporal gyrus). This approach requires a gyrectomy of the fusiform gyrus to obtain access to the mesial structures. Hori, et al. also demonstrated in the inset figure that incising and reflecting the tentorium benefits accessing the mesial temporal structures for removal.

Over many years now, a transcortical approach through the second temporal gyrus has been used by the author similar to the approach originally described by Niemeyer [58,59]. SAH achieves to provide the best possible outcomes of seizure freedom by limiting the resection to the epileptic focus and sparing from resection normal structures that are not primarily a part of the seizure focus. At centers with experience in SAH, the results on seizure freedom are the same as the larger anterior temporal resection or cortico-amygdalohippocampectomy (CAH) [24–29,35,38,44,56,58,60]. The neuropsychological results of SAH appear to be improved compared to CAH although this conclusion is not shared by all investigations on the topic [42,61,62]. However, the objective of SAH to spare brain tissue not involved with the seizure focus is an important goal of epilepsy surgery in general.

3. Surgical Technique of SAH

The scalp and bone exposure for SAH can range from a frontotemporal craniotomy to a keyhole minimal access approach. Electrocorticography (ECOG) requires the larger craniotomy to fit the surface electrodes in the bony opening. However, if ECOG is not needed, keyhole minimal access is the preferred approach.

The traditional frontotemporal craniotomy is exposed via a curvilinear "?" mark scalp incision that starts at the zygoma, extends posterior just to the back of the ear, up to the superior temporal line, and anterior to the hair line. The temporalis muscle is reflected anteriorly with scalp. Exposing the root of the zygoma is important to gain adequate exposure to the floor of the middle fossa. After elevation of the frontotemporal bone flap, a craniectomy of the temporal bone is required to extend the bone opening to the zygoma inferiorly and to the anterior extent of the middle fossa. After the dura is reflected away from the temporal lobe, important sulcal and gyral landmarks, such as the Sylvian fissure, superior temporal gyrus (T1), and the central sulcus are confirmed with image guidance.

A corticectomy along the superior aspect of the second temporal gyrus (T2) is made to begin the corridor to the mesial temporal structures. A 2 to 3 cm corticectomy provides sufficient working space. Figure 5 The posterior extent of the corticectomy is limited by the central sulcus in the non-dominant hemisphere and by the precentral sulcus in the dominant hemisphere. Image guidance is sufficient to localize precentral and central sulcus, which corresponds to about 3.5 cm and 4.5 cm from the temporal tip, respectively. The author has found that using an ultrasonic aspirator at the lowest settings of aspiration and amplitude to be the best subpial dissection and aspiration tool. A corridor is fashioned down to the temporal horn that is opened, which is typically 3–4 cm deep from the cortical surface. The corridor trajectory parallels the superior temporal sulcus, which points to the temporal horn and stops just short of it. A retractor is inserted once the temporal horn is opened to facilitate fully exposing the mesial temporal structures by opening the ventricle from its anterior most tip to the posterior limit of the corridor Figure 6.

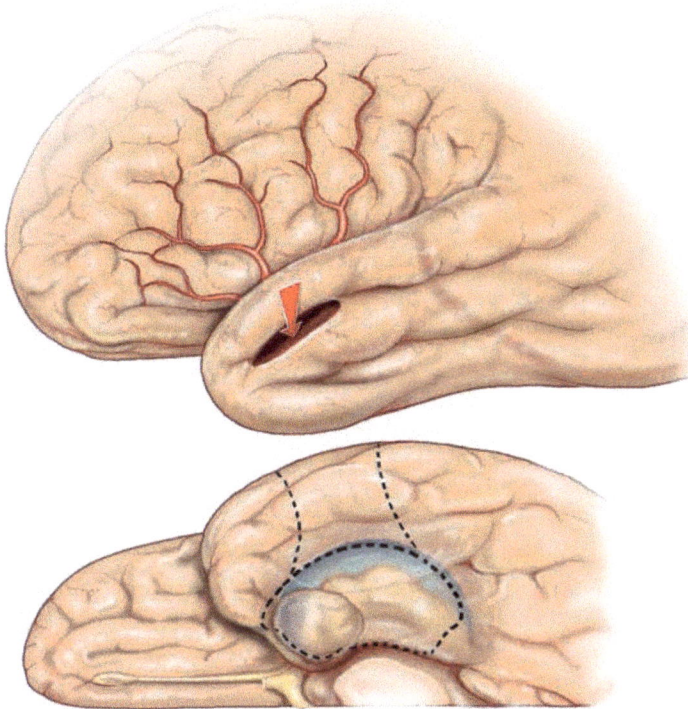

Figure 5. SAH corticectomy of about 2.5 cm is made along T2 just below the superior temporal sulcus. A white matter corridor is fashioned that follows the superior temporal sulcus down to the temporal horn, which is opened to visualize directly the mesial temporal structures.

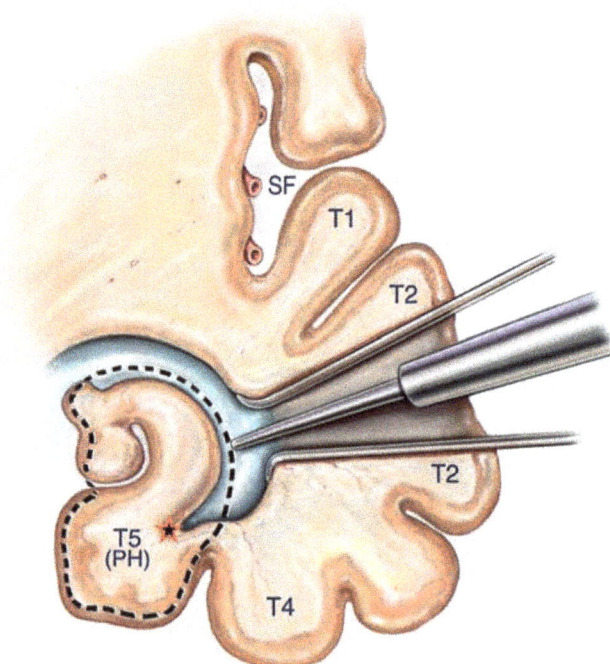

Figure 6. Reprinted from Journal of Clinical Neuroscience 17 (9), Boling W, Minimal access keyhole surgery for mesial temporal lobe epilepsy, 1180–1184, Copyright 2010, with permission from Elsevier, [59]. Coronal view of the temporal lobe and nearby structures. A retractor is placed along the white matter corridor after opening the ventricle to fully expose its contents. The first step in the SAH is to enter the lateral ventricular sulcus (star) that lies between the bulges into the ventricle of the hippocampus and collateral eminence in order to empty the parahippocampus in a subpial fashion. Dotted line illustrates mesial temporal structures to be removed in this view, namely the hippocampal complex and parahippocampus. SF = Sylvian Fissure, T1= superior temporal gyrus, T2 = middle temporal gyrus, T4 = Fusiform gyrus, T5(PH) = parahippocampal gyrus.

The resection of the mesial temporal structures is subpial preferably using the ultrasonic aspirator at its lowest settings of aspiration and amplitude, which is an excellent tool for subpial resection. There are several key intraventricular landmarks identified in the mesial temporal region that guide the SAH surgery. The lateral ventricular sulcus lying between the two bulges into the ventricle of the hippocampus and the collateral eminence (see Figures 6 and 7) is the entry into the parahippocampal gyrus that is emptied subpially as the first surgical maneuver. The P2 segment of the posterior cerebral artery is usually identified bulging into the parahippocampus, which fills the space between the collateral sulcus and hippocampal sulcus. The hippocampus can now be retracted laterally into the space created by removing the parahippocampus in order to aspirate the fimbria from the underlying pia. The hippocampus is next removed en bloc by disconnection from the hippocampal tail posteriorly, which enables the hippocampus to be elevated and separated from its vascular pedicle, the hippocampal sulcus, by teasing and separating vessels from the hippocampus or by coagulation and division. Additional hippocampus can be aspirated and removed back to the typical posterior limit, which corresponds to the level of the midbrain tectum, a landmark visualized readily with image guidance.

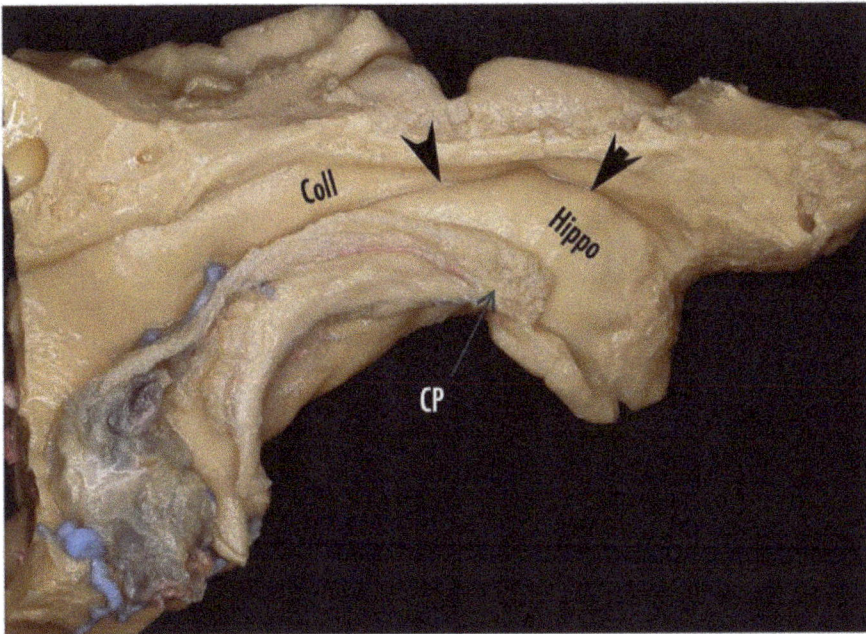

Figure 7. Cadaver dissection of the mesial temporal lobe cut longitudinally along the temporal horn to illustrate the mesial temporal structures. Coll = collateral eminence, arrowheads point to the lateral ventricular sulcus, Hippo = head of the hippocampus, CP = choroid plexus.

The anterior extent of the SAH resection includes the entire uncus that should be completely emptied subpial. The uncus is composed mostly of the hippocampus and dentate gyri that have curved back on themselves to create a hook-like appearance. In very close proximity to the transparent pia of the uncus, cranial nerve III and the P1 segment of the posterior cerebral artery are typically visualized beneath the pia lying in their cistern. Figure 8 The amygdala, which contributes to the dorsomedial portion of the uncus, called the semilunar gyrus, is either aspirated completely or dissected around and removed en bloc. The choroid plexus is an important landmark to identify early and repeatedly refer to because it limits the mesiosuperior resection extent.

The completion of the SAH corresponds to removal of the mesial temporal structures found between the choroid plexus and collateral sulcus that includes the uncus, amygdala, hippocampus, parahippocampus, and fimbria. All of which are removed by keeping the underlying pia intact. The anterior limit is to the anterior extent of the uncus at the tentorial incisura and the posterior limit is back to the level of the midbrain tectum (Figure 9).

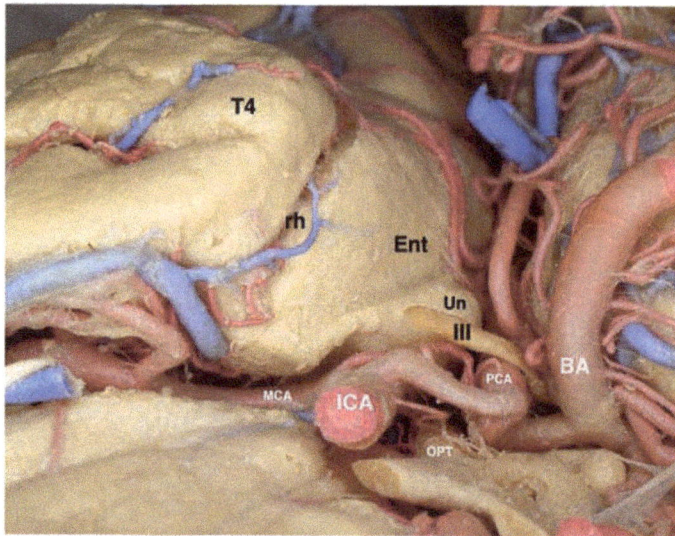

Figure 8. View of a fixed and injected brain from the orbital frontal surface looking posteriorly. The normal relationships of structures lying adjacent to the mesial temporal lobe are visualized. The 3rd cranial nerve is normally abutting the pia of the anterior uncus. The posterior cerebral artery can be recognized along its course beneath the transparent pia of the uncus and the parahippocampus. The close association of the ICA and its bifurcation with the uncus and amygdala are illustrated. T4 = fusiform gyrus of the temporal lobe, rh = rhinal sulcus, Ent = entorhinal cortex (most anterior extent of the parahippocampus), Un = uncus, III = oculomotor cranial nerve, ICA = internal carotid artery, PCA = posterior cerebral artery, OPT = optic chiasm, BA = basilar artery, MCA = middle cerebral artery.

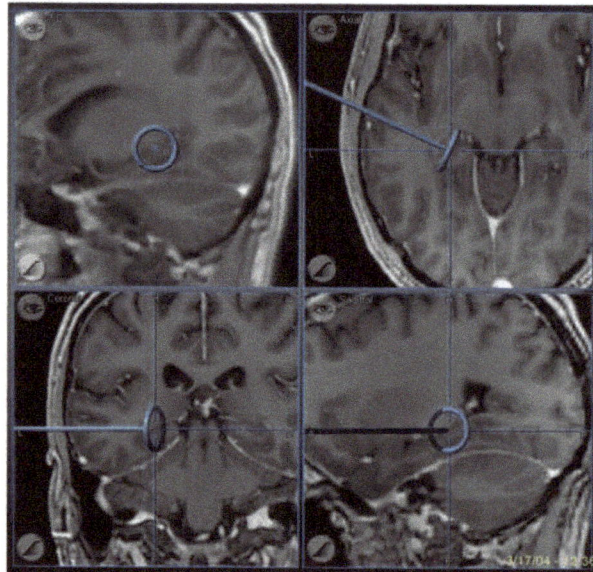

Figure 9. Image guidance view of the navigation pointer at the posterior extent of the hippocampal removal, which corresponds to the level of the midbrain tectum.

4. Key Hole Approach in SAH

Minimal access surgery has potential to benefit patients by providing more rapid recovery and improved patient outcomes compared with traditional larger surgical incisions and exposures [60,63,64]. The minimal access (or keyhole) approach lends itself quite well to SAH using either the trans-middle temporal gyrus or subtemporal approaches since the necessary maneuvers required to remove the mesial temporal structures can be accomplished with relative ease through a small scalp incision and bony opening [29,59,60,65]. The SAH approach was envisioned and developed in order to spare from resection cerebral tissue that is not a part of the primary seizure focus. Therefore, minimizing the scalp and bony opening to accomplish SAH represents a logical progression from the larger more traditional scalp incision and craniotomy exposure. In individuals with clear-cut MTLE who are candidates for SAH, the keyhole, minimal access approach is preferred. Because the exposure is too limited to confidently recognize sulcal and gyral landmarks, keyhole SAH should be performed with the aid of image guidance.

The scalp incision is curvilinear starting at the zygoma. Figure 10 The temporalis muscle is split and held open with hooks or a self-retaining retractor. The bone opening is a silver dollar sized craniotomy plus additional craniectomy anterior and inferior centered over the second temporal gyrus and the planned cortical incision. Figure 11 The keyhole and non-keyhole SAH approach to the mesial temporal structures are essentially identical. Image guidance is critical to confirm the corticectomy is positioned properly in the middle temporal gyrus anterior to the central sulcus in the non-dominant hemisphere and anterior to the pre-central sulcus in the dominant hemisphere.

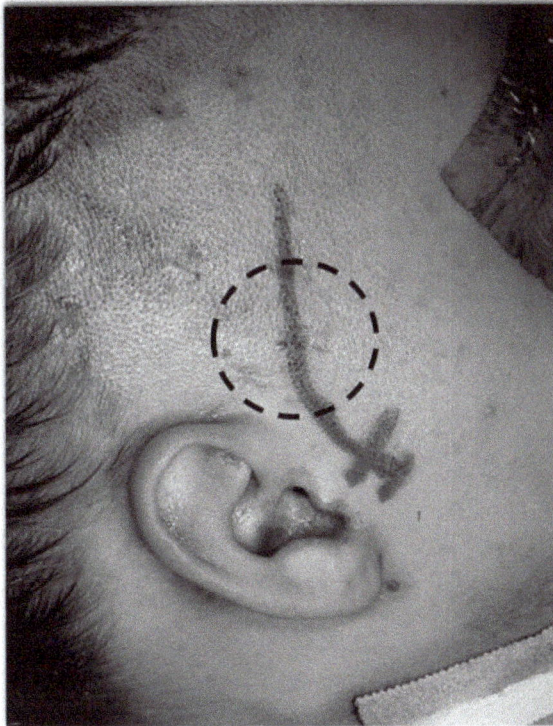

Figure 10. The keyhole minimal access approach benefits the patient with smaller skin incision and cranial opening. A slight curve in the scalp incision helps exposure and reduces retraction forces on the temporalis muscle compared with a straight linear incision.

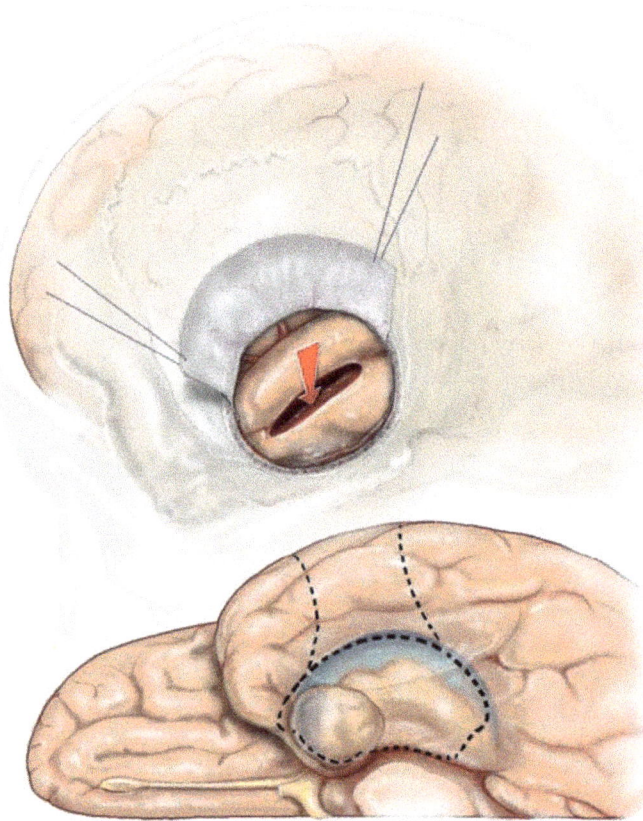

Figure 11. Reprinted from Journal of Clinical Neuroscience 17 (9), Boling W, Minimal access keyhole surgery for mesial temporal lobe epilepsy, 1180–1184, Copyright 2010, with permission from Elsevier, [59]. The corticectomy and trans-T2 trans-ventricular approach to resection of the mesial temporal structures is identical to the approach via a standard scalp incision and cranial opening. The exposure accomplished with a keyhole access approach is more than adequate to perform the maneuvers required to complete the SAH.

5. Percutaneous Ablation Approaches in SAH

For the past many decades there has been interest in the stereotactic insertion of a probe to target the mesial temporal structures for thermal ablation in patients with TLE. The early efforts used stereotactic radiofrequency thermal ablation techniques that were also being used at the time to lesion the thalamus to treat movement disorder, the Gasserian ganglion to eliminate the pain of trigeminal neuralgia, and other neurological disorders that could be treated stereotactically. The amygdala was solely targeted in the initial reports of stereotactic ablation in the treatment of epilepsy. The reports of Schwab, et al. [66] and Narabayashi, et al. [67] described approximately 30% seizure free patients after stereotactic amygdalectomy for TLE. Flanigin and Nashold, both former fellows of the Montreal Neurological Institute, were able to demonstrate the feasibility of multiple stereotactic ablations to expand the volume of mesial temporal structures treated to include the hippocampus along with the amygdala, as well as bilateral amygdalectomies [68]. However, more reports of amygdalectomy and hippocampectomy treated with stereotactic radiofrequency probes continued to be disappointing in regards to seizure free rates and did not approach the seizure free successes achieved with anterior temporal lobe resection [69–72]. The early efforts at stereotactic ablation of the mesial temporal structures to treat intractable TLE were largely hampered by the challenges of targeting the mesial temporal structures using the methods of the time; stereotactic targets were defined by imaging with angiogram and pneumoencephaolgraphy or by measurements made from the AC-PC line. Also, the early stereotactic frames were best suited for orthogonal trajectories, which made treatment along the longitudinal axis of the mesial temporal structures a more arduous treatment plan.

With the advent of CT and MR imaging, the mesial temporal structures could be directly and more accurately targeted for stereotactic treatment. Parrent and Blume reported the experience of stereotactic amygdalohippocampectomy for TLE with the use of modern MR imaging [73]. The authors described outcomes from patients with TLE and evidence of exclusively ipsilateral MTS in 11 out of 19 patients treated. Surgery was performed using a stereotactic headframe with radiofrequency lesioning of the amygdala and hippocampus. Postoperative MRI identified the best results were in patients with more complete stereotactic ablations involving the entire amygdala and measuring 15–34 mm along the hippocampus. In this group of patients, 9 out of 15 had favorable results defined as greater than 90% reduction in seizure frequency or seizure freedom. Only 1 out 5 patients with incomplete ablation of the amygdala and hippocampus had a favorable result. Parrent and Blume were able to demonstrate the improved results of stereotactic ablation surgery using modern imaging to target the thermal probe, and that more complete treatment of the amygdala and hippocampus produced better results of seizure freedom and seizure frequency reduction.

Three additional advances have enabled further improvement in the technique of stereotactic ablation surgery of TLE. One is frameless stereotaxy or image guidance. Now that stereotactic surgery can be performed without the traditional stereotactic headframe in place, neurosurgeons have considerably more freedom to plan non-orthogonal trajectories to targets that were more difficult or even impossible to accomplish previously. Another advance is the development of low cost fiber-optic lasers for thermal ablation. The radiofrequency ablation probe has not been able to be manufactured to allow use in the MRI magnet. Using the laser ablation technology, only the non-ferromagnetic fiber optic cable is in the MRI environment. Therefore, the thermal treatment can be delivered with the patient in the MRI to confirm the anatomy treated and the thermal dose delivered, which segues to the third advance, the ability to create a thermal map using normal MRI acquisition protocols. Several measurable MRI variables are affected by temperature in a linear relationship and can be used to construct a thermal map, such as diffusion coefficient [74], chemical shift, T1 relaxation [75], magnetization transfer, proton density and proton resonance frequency (PRF). The excellent linearity and temperature dependence of the PRF in relation to most all tissue types have made PRF-based phase mapping methods the preferred choice for thermal mapping and this acquisition protocol is available on most all modern MRI units.

There are 2 FDA approved laser thermal ablation devices that overlay anatomy with a thermal map, NeuroBlate (Monteris Medical, Plymouth, MN) and Visualase (Medtronic, Minneapolis, MN). Both devices use a pulsed laser thermal source. The probe, which consists of a fiber-optic cable to deliver the laser plus a cooling system, is inserted stereotactically to the desired target in the operating room. Then the actual thermal ablation treatment is performed in the MRI magnet. There has been considerable renewed interest in the use of thermal ablation for the treatment of intractable MTLE, and the fiber-optic laser technologies have improved the ability to target the mesial temporal structures and assess the treatment delivered essentially real-time.

The fiber-optic laser thermal probe is typically inserted in a posterior to anterior trajectory to allow treatment of the mesial temporal structures along the longitudinal axis. Multiple thermal treatments are delivered as the fiber-optic probe is pulled back until a sufficient volume of mesial temporal structures are ablated. One of the challenges of treatment planning is to insert the probe in such a trajectory that the uncus and hippocampal head can be completely ablated. Jermakowicz, et al. found lateral trajectories that did not include the uncus and hippocampal head in the ablation treatment were significantly correlated with persistent disabling seizures [76]. Overall the laser ablation of mesial temporal structures has been shown to be a safe minimally invasive alternative to craniotomy surgery of MTLE. Although the data thus far point to reduced odds of seizure freedom in patients treated with laser thermal ablation versus craniotomy approaches to resect the mesial temporal structures [77].

6. Patient Outcomes from SAH

The success of SAH to achieve seizure freedom is dependent on the accurate diagnosis of MTLE lateralized to the side of surgery and the successful removal or disconnection of the seizure focus at surgery. At centers with expertise in the diagnosis of MTLE, the seizure freedom rates after SAH are the same as anterior temporal resection [29,35,59,60,65,78,79]. To date a randomized controlled study to evaluate and compare the surgical approaches of SAH and ATL for the treatment of MTLE has not been undertaken. The literature up to now demonstrates no clear difference in ability to become seizure free in patients with surgery of SAH or ATL for MTLE (Table 1).

Table 1. Studies comparing ATL with SAH for outcomes of seizure freedom and reducing seizure frequency.

Renowden et al. (1995) [80]	Same outcome for two types of SAH
Arruda et al. (1996) [81]	Similar
Pauli et al. (1999) [82]	Similar
Clusmann et al. (2002) [83]	Similar
Clusmann et al. (2004) [40]	Better in ATL in children + adolescents
Lutz et al. (2004) [40]	Same outcome for two types of SAH
Paglioli et al. (2006) [84]	Similar
Tanriverdi, et al. (2007) [85]	Similar
Bate et al. (2007) [86]	Better in ATL in children + adolescents
Tanriverdi, et al. (2010) [87]	Similar
Sagher, et al. (2012) [78]	Similar
Wendling, et al. (2013) [88]	Similar
Bujarski, et al. (2013) [89]	Similar
Josephson, et al. (2013) [90]	Better outcome in ATL
Hu, et al. (2013) [91]	Better outcome in ATL
Nascimento, et al. (2016) [92]	Similar although ATL associated with more complications
Schmeiser, et al. (2017) [93]	No difference in three types of SAH nor in SAH versus ATL
Foged, et al. (2018) [94]	No difference at 1 year and 7 years after surgery

One potential advantage of SAH is that cognitive function may be improved in patients with more limited temporal lobe resections. This topic was reviewed by Schramm [94] who found the literature lacking consistency on reporting of crucial study variables such as MRI confirmation of extent of resection and studies were mostly retrospective reports. However, Schramm concluded from the available literature that "in summary there is considerable evidence for somewhat better neuropsychological results with SAH, although this was not consistently found." An analysis of studies published from 1992 to 2018 supports Schramm's conclusion that in general there tends to be more neuropsychological deficits in ATL than SAH surgery, although the heterogeneity of studies and the lack of a randomized controlled study comparing the two surgical techniques makes firm conclusions about neuropsychological impact impossible to make (Table 2).

Table 2. Studies comparing temporal lobe surgery of ATL with SAH for neuropsychological outcome.

Goldstein, et al. (1992) [95]	No difference when using global memory test
Goldstein, et al. (1993) [96]	SAH short-term beneficial effect on memory
Wolf et al. (1993) [97]	No difference
Renowden et al. (1995) [80]	SAH better in verbal IQ and nonverbal memory
Helmstaedter et al. (1996) [98]	Immediate recall better in SAH
Jones-Gotman (1997) [42]	Similar deficits in learning + retention tasks in seizure free patients
Pauli et al. (1999) [82]	SAH better for verbal memory
Helmstaedter et al. (2002) [99]	SAH has advantage over ATL in long-term follow-up (2–10 years)
Clusmann (2004) [40]	SAH in adults: higher rate of improvement + lower rate of deterioration in overall neuropsychological score
Hader et al. (2005) [100]	No difference
Morino et al. (2006) [101]	SAH better memory function
Paglioli et al. (2006) [84]	SAH better for verbal memory score (30% deterioration in both groups)
Tanriverdi, et al. (2007) [85]	SAH less decline for verbal memory
Helmstaedter et al. (2008) [102]	SAH better for R-sided resection, ATL better for L-sided resection for material-specific memory
Tanriverdi, et al. (2010) [87]	Worse verbal IQ after SAH
Sagher, et al. (2012) [78]	No difference
Bujarski, et al. (2013) [89]	No difference, except more post-surgical paranoia after ATL
Wendling, et al. (2013) [88]	Worse memory after ATL
Boucher, et al. (2015) [103]	Worse after ATL on immediate recall of Logical Memory subtest of Wechsler Memory Scales. Delayed recognition trial of Rey Auditory Verbal Learning Test worse after SAH
Nascimento, et al. (2016) [92]	No difference
Gül, et al. (2017) [104]	No difference
Schmeiser, et al. (2017) [93]	No difference
Foged, et al. (2018) [94]	SAH better than ATL verbal memory in left hemisphere surgery only

An additional outcome parameter of considerable importance to the patient is the ease and speed of recovery from brain surgery. SAH has an opportunity to provide patients shorter operative time, quicker recovery, and reduced length of stay in the hospital particularly if surgery is performed using minimal access or key hole approaches [59,105]. Reduced hospital length of stay is a marker of both quicker patient recovery and a more cost effective operation. Boling identified significantly reduced operative time and shorter hospital length of stay in patients who underwent key hole craniotomy versus a standard craniotomy approach for SAH [59].

7. Conclusions

Patients with medically intractable MTLE have an excellent opportunity for seizure freedom from resection surgery that can be a standard anterior temporal removal or selective amygdalohippocampectomy. The potential advantage of the SAH is a selective removal of the seizure focus sparing temporal lobe regions that are not the actual epileptogenic zone. SAH, especially minimal access approaches, provide direct advantages of quicker patient recovery from surgery, and evidence points to a probable cognitive benefit compared with standard temporal resections.

Conflicts of Interest: The author reports no conflicts of interest.

References

1. Téllez-Zenteno, J.F.; Hernández-Ronquillo, L. A Review of the Epidemiology of Temporal Lobe Epilepsy. *Epilepsy Res. Treat.* **2012**, 630853. [CrossRef] [PubMed]
2. Semah, F.; Picot, M.C.; Adam, C.; Broglin, D.; Arzimanoglou, A.; Bazin, B.; Cavalcanti, D.; Baulac, M. Is the underlying cause of epilepsy a major prognostic factor for recurrence? *Neurology* **1998**, *51*, 1256–1262. [CrossRef] [PubMed]
3. Hauser, W.A.; Annegers, J.F.; Kurland, L.T. Prevalence of epilepsy in Rochester, Minnesota: 1940–1980. *Epilepsia* **1991**, *32*, 429–445. [CrossRef] [PubMed]
4. Papez, J.W. A proposed mechanism of emotion. *Arch. Neurol. Psychiatry* **1937**, *38*, 725–743. [CrossRef]
5. Duvernoy, H.M.; Cattin, F.; Risold, P.Y. *The Human Hippocampus Functional Anatomy, Vascularization and Serial Sections with MRI*; Springer: New York, NY, USA, 2013.
6. Wiebe, S.; Blume, W.T.; Girvin, J.P.; Eliasziw, M. A randomized, controlled trial of surgery for temporal-lobe epilepsy. *N. Engl. J. Med.* **2001**, *345*, 311–318. [CrossRef] [PubMed]
7. Tanriverdi, T.; Ajlan, A.; Poulin, N.; Olivier, A. Morbidity in epilepsy surgery: An experience based on based on 2449 epilepsy surgery procedures from a single institution. *J. Neurosurg.* **2009**, *110*, 1111–1123. [CrossRef] [PubMed]
8. Engel, J., Jr.; Wiebe, S.; French, J.; Sperling, M.; Williamson, P.; Spencer, D.; Gumnit, R.; Zahn, C.; Westbrook, E.; Enos, B. Practice parameter: Temporal lobe and localized neocortical resections for epilepsy: Report of the Quality Standards Subcommittee of the American Academy of Neurology, in association with the American Epilepsy Society and the American Association of Neurological Surgeons. *Neurology* **2003**, *60*, 538–547.
9. Kwan, P.; Arzimanoglou, A.; Berg, A.T.; Brodie, M.J.; Allen Hauser, W.; Mathern, G.; Moshé, S.L.; Perucca, E.; Wiebe, S.; French, J. Definition of drug resistant epilepsy: Consensus proposal by the ad hoc Task Force of the ILAE Commission on Therapeutic Strategies. *Epilepsia* **2010**, *51*, 1069–1077. [CrossRef] [PubMed]
10. Kwon, C.S.; Neal, J.; Telléz-Zenteno, J.; Metcalfe, A.; Fitzgerald, K.; Hernandez-Ronquillo, L.; Hader, W.; Wiebe, S.; Jetté, N.; CASES Investigators. Resective focal epilepsy surgery—Has selection of candidates changed? A systematic review. *Epilepsy Res.* **2016**, *122*, 37–43. [CrossRef] [PubMed]
11. Wieser, H.G.; ILAE Commission on Neurosurgery of Epilepsy. Mesial temporal lobe epilepsy with hippocampal sclerosis. *Epilepsia* **2004**, *45*, 695–714. [PubMed]
12. O'Brien, T.J.; Kilpatrick, C.; Murrie, V.; Vogrin, S.; Morris, K.; Cook, M.J. Temporal lobe epilepsy caused by mesial temporal sclerosis and temporal neocortical lesions. A clinical and electroencephalographic study of 46 pathologically proven cases. *Brain* **1996**, *119*, 2133–2141. [CrossRef] [PubMed]
13. Burgerman, R.S.; Sperling, M.R.; French, J.A.; Saykin, A.J.; O'Connor, M.J. Comparison of mesial versus neocortical onset temporal lobe seizures: Neurodiagnostic findings and surgical outcome. *Epilepsia* **1995**, *36*, 662–670. [CrossRef] [PubMed]
14. Olivier, A.; Gloor, P.; Andermann, F.; Quesney, L.F. The place of stereotactic depth electrode recording in epilepsy. *Appl. Neurophysiol.* **1985**, *48*, 395–399. [CrossRef] [PubMed]
15. Berkovic, S.F.; Andermann, F.; Olivier, A.; Ethier, R.; Melanson, D.; Robitaille, Y.; Kuzniecky, R.; Peters, T.; Feindel, W. Hippocampal sclerosis in temporal lobe epilepsy demonstrated by magnetic resonance imaging. *Ann. Neurol.* **1991**, *29*, 175–182. [CrossRef] [PubMed]
16. Cendes, F. Febrile seizures and mesial temporal sclerosis. *Curr. Opin. Neurol.* **2004**, *17*, 161–164. [CrossRef] [PubMed]
17. Schramm, J.; Kral, T.; Grunwald, T.; Blumcke, I. Surgical treatment for neocortical temporal lobe epilepsy: Clinical and surgical aspects and seizure outcome. *J. Neurosurg.* **2001**, *94*, 33–42. [CrossRef] [PubMed]
18. Pacia, S.V.; Devinsky, O.; Perrine, K.; Ravdin, L.; Luciano, D.; Vazquez, B.; Doyle, W.K. Clinical features of neocortical temporal lobe epilepsy. *Ann. Neurol.* **1996**, *40*, 724–730. [CrossRef] [PubMed]
19. Maillard, L.; Vignal, J.P.; Gavaret, M.; Guye, M.; Biraben, A.; McGonigal, A.; Chauvel, P.; Bartolomei, F. Semiologic and electrophysiologic correlations in temporal lobe seizure subtypes. *Epilepsia* **2004**, *45*, 1590–1599. [CrossRef] [PubMed]
20. Adam, C.; Clemenceau, S.; Semah, F.; Hasboun, D.; Samson, S.; Aboujaoude, N.; Samson, Y.; Baulac, M. Variability of presentation in medial temporal lobe epilepsy: A study of 30 operated cases. *Acta. Neurol. Scand.* **1996**, *94*, 1–11. [CrossRef] [PubMed]

21. Pfander, M.; Arnold, S.; Henkel, A.; Weil, S.; Werhahn, K.J.; Eisensehr, I.; Winkler, P.A.; Noachtar, S. Clinical features and EEG findings differentiating mesial from neocortical temporal lobe epilepsy. *Epileptic Disord.* **2002**, *4*, 189–195. [PubMed]

22. Foldvary, N.; Lee, N.; Thwaites, G.; Mascha, E.; Hammel, J.; Kim, H.; Friedman, A.H.; Radtke, R.A. Clinical and electrographic manifestations of lesional neocortical temporal lobe epilepsy. *Neurology* **1997**, *49*, 757–763. [CrossRef] [PubMed]

23. So, N.; Gloor, P.; Quesney, L.F.; Jones-Gotman, M.; Olivier, A.; Andermann, F. Depth electrode investigations in patients with bitemporal epileptiform abnormalities. *Ann. Neurol.* **1989**, *25*, 423–431. [CrossRef] [PubMed]

24. Olivier, A.; Boling, W.; Tanriverdi, T. (Eds.) *Techniques in Epilepsy Surgery: The MNI Approach*; Cambridge University Press: New York, NY, USA, 2012.

25. Spencer, D.D.; Spencer, S.S.; Mattson, R.H.; Williamson, P.D.; Novelly, R.A. Access to the posterior medial temporal lobe structures in the surgical treatment of temporal lobe epilepsy. *Neurosurgery* **1984**, *15*, 667–671. [CrossRef] [PubMed]

26. Olivier, A. Temporal resections in the surgical treatment of epilepsy. *Epilepsy Res. Suppl.* **1992**, *5*, 175–188. [PubMed]

27. Silbergeld, D.L.; Ojemann, G.A. The tailored temporal lobectomy. *Neurosurg. Clin. N. Am.* **1993**, *4*, 273–281. [PubMed]

28. Kim, H.I.; Olivier, A.; Jones-Gotman, M.; Primrose, D.; Andermann, F. Corticoamygdalectomy in memory-impaired patients. *Stereotact. Funct. Neurosurg.* **1992**, *58*, 162–167. [CrossRef] [PubMed]

29. Boling, W.; Longoni, N.; Palade, A.; Moran, M.; Brick, J. Surgery for temporal lobe epilepsy. *W V Med. J.* **2006**, *102*, 18–21. [PubMed]

30. Jackson, J.H.; Colman, W.S. Case of epilepsy with tasting movements and "dreamy state" with very small patch of softening in the left uncinate gyrus. *Brain* **1898**, *21*, 580–590. [CrossRef]

31. Kaada, B.R. Somatomotor, autonomic and electrographic responses to electrical stimulation of "rhinencephalic" and other structures in primates cat and dog: A study of responses from limbic, subcallosal, orbito-insular, pyriform and temporal cortex, hippocampus, fornix and amygdala. *Acta. Physiol. Scand. Suppl.* **1951**, *83*, 1–285.

32. Vigouroux, R.; Gastaut, H.R.; Badier, M. Provocation des principales manifestations cliniques de l'épilepsie dite temporale par stimulation des structures rhinencéphaliques chez le chat non-anaesthésié. *Rev. Neurol.* **1951**, *85*, 505–508. [PubMed]

33. Gastaut, H.R.; Vigouroux, R.; Naquet, R. Lésions épileptogènes amygdalo-hippocampiques provoquées chez le chat par l'injection de "crème d'albumine". *Rev. Neurol.* **1952**, *87*, 607–609. [PubMed]

34. Green, J.D.; Shimamoto, T. Hippocampal seizures and their propagation. *Arch. Neurol. Psychiatry* **1953**, *7*, 687–702. [CrossRef]

35. Wieser, H.G.; Ortega, M.; Friedman, A.; Yonekawa, Y. Long-term seizure outcomes following amygdalohippocampectomy. *J. Neurosurg.* **2003**, *98*, 751–763. [CrossRef] [PubMed]

36. Penfield, W.; Jasper, H. *Epilepsy and the Functional Anatomy of the Human Brain*, 1st ed.; Little Brown: Boston, MA, USA, 1954.

37. Olivier, A. Relevance of removal of limbic structures in surgery for temporal lobe epilepsy. *Can. J. Neurol. Sci.* **1991**, *18* (Suppl. 4), 628–635. [CrossRef] [PubMed]

38. Olivier, A. Transcortical selective amygdalohippocampectomy in temporal lobe epilepsy. *Can. J. Neurol. Sci.* **2000**, *27* (Suppl. 1), S68–S76. [CrossRef] [PubMed]

39. Abosch, A.; Bernasconi, N.; Boling, W.; Jones-Gotman, M.; Poulin, N.; Dubeau, F.; Andermann, F.; Olivier, A. Factors predictive of suboptimal seizure control following selective amygdalohippocampectomy. *J. Neurosurg.* **2002**, *97*, 1142–1151. [CrossRef] [PubMed]

40. Lutz, M.T.; Clusmann, H.; Elger, C.E.; Schramm, J.; Helmstaedter, C. Neuropsychological outcome after selective amygdalohippocampectomy with transsylvian versus transcortical approach: A randomized prospective clinical trial of surgery for temporal lobe epilepsy. *Epilepsia* **2004**, *45*, 809–816. [CrossRef] [PubMed]

41. Niemeyer, P. The Transventricular Amygdala-Hippocampectomy in Temporal Lobe Epilepsy. In *Temporal Lobe Epilepsy: A Colloquium Sponsored by the National Institute of Neurological Diseases and Blindness, National Institutes of Health, Bethesda, Maryland, in Cooperation with the International League Against Epilepsy*; Baldwin, M., Bailey, P., Eds.; Charles C Thomas: Springfield, IL, USA, 1958; pp. 461–482.

42. Jones-Gotman, M.; Zatorre, R.J.; Olivier, A.; Andermann, F.; Cendes, F.; Staunton, H.; McMackin, D.; Siegel, A.M.; Wieser, H.G. Learning and retention of words and designs following excision from medial or lateral temporal-lobe structures. *Neuropsychologia* **1997**, *35*, 963–973. [CrossRef]

43. Wieser, H.G. Selective amygdalo-hippocampectomy for temporal lobe epilepsy. *Epilepsia* **1988**, *29* (Suppl. 2), S100–S113. [CrossRef] [PubMed]

44. Tanriverdi, T.; Olivier, A.; Poulin, N.; Andermann, F.; Dubeau, F. Long-term seizure outcome after mesial temporal lobe epilepsy surgery: Corticalamygdalohippocampectomy versus selective amygdalohippocampectomy. *J. Neurosurg.* **2008**, *108*, 517–524. [CrossRef] [PubMed]

45. JAMA Network. Temporal Lobe Epilepsy: A Colloquium Sponsored by the National Institute of Neurological Diseases and Blindness, National Institutes of Health, Bethesda, Maryland, in Cooperation with the International League Against Epilepsy. *JAMA* **1959**, *169*, 1143–1144. [CrossRef]

46. Scoville, W.; Milner, B. Loss of recent memory after bilateral hippocampal lesions. *J. Neurol. Neurosurg. Psych.* **1957**, *20*, 11–21. [CrossRef]

47. Penfield, W.; Milner, B. Memory deficit produced by bilateral lesions in the hippocampal zone. *Arch. Neurol. Psychiatry* **1958**, *79*, 475–497. [CrossRef]

48. Penfield, W.; Flanigin, H. Surgical therapy of temporal lobe seizures. *Arch. Neurol. Psychiatry* **1950**, *64*, 491–500. [CrossRef] [PubMed]

49. Gastaut, H. So-called "psychomotor" and "temporal" epilepsy. *Epilepsia* **1953**, *2*, 59–76. [CrossRef]

50. Sano, K.; Malamud, N. Clinical significance of sclerosis of the cornu ammonis. *Arch. Neurol. Psychiatry* **1953**, *70*, 40–53. [CrossRef]

51. Hill, D.; Falconer, M.A.; Pampiglione, G.; Liddell, D.W. Discussion on the surgery of temporal lobe epilepsy. *Proc. R Soc. Med.* **1953**, *46*, 965–976. [PubMed]

52. Feindel, W.; Penfield, W. Localization of discharge in temporal lobe automatism. *Arch. Neurol. Psychiatry* **1954**, *72*, 605–630. [CrossRef]

53. Yasargil, M.G.; Teddy, P.J.; Roth, P. Selective amygdalo-hippocampectomy: Operative anatomy and surgical technique. In *Advances and Technical Standards in Neurosurgery*; Symon, L., Ed.; Springler-Wien: New York, NY, USA, 1985; Volume 12.

54. Martens, T.; Merkel, M.; Holst, B.; Brückner, K.; Lindenau, M.; Stodieck, S.; Fiehler, J.; Westphal, M.; Heese, O. Vascular events after transsylvian selective amygdalohippocampectomy and impact on epilepsy outcome. *Epilepsia* **2014**, *55*, 763–769. [CrossRef] [PubMed]

55. Yasargil, M.G.; Teddy, P.J.; Roth, P. Selective Amygdalo-Hippocampectomy Operative Anatomy and Surgical Technique. *Adv. Tech. Stand. Neurosurg.* **1985**, *12*, 93–123. [PubMed]

56. Hori, T.; Tabuchi, S.; Kurosaki, M.; Kondo, S.; Takenobu, A.; Watanabe, T. Subtemporal amygdalohippocampectomy for treating medially intractable temporal lobe epilepsy. *Neurosurgery* **1993**, *33*, 50–56. [PubMed]

57. Sugita, K.; Kobayashi, S.; Yokoo, A. Preservation of large bridging veins during brain retraction. Technical note. *J. Neurosurg.* **1982**, *57*, 856–858. [CrossRef] [PubMed]

58. Boling, W. Selective Amygdalohippocampectomy. In *Operative Techniques in Epilepsy Surgery*, 2nd ed.; Baltuch, G., Cukiert, A., Eds.; Thieme Medical Publishers: New York, NY, USA, 2018.

59. Boling, W. Minimal Access Keyhole Surgery for Mesial Temporal Lobe Epilepsy. *J. Clin. Neurosci.* **2010**, *17*, 1180–1184. [CrossRef] [PubMed]

60. Little, A.S.; Smith, K.A.; Kirlin, K.; Baxter, L.C.; Chung, S.; Maganti, R.; Treiman, D.M. Modifications to the subtemporal selective amygdalohippocampectomy using a minimal-access technique: Seizure and neuropsychological outcomes. *J. Neurosurg.* **2009**, *111*, 1263–12674. [CrossRef] [PubMed]

61. Kessels, R.P.; Hendriks, M.; Schouten, J.; Van Asselen, M.; Postma, A. Spatial memory deficits in patients after unilateral selective amygdalohippocampectomy. *J. Int. Neuropsychol. Soc.* **2004**, *10*, 907–912. [CrossRef] [PubMed]

62. Gleissner, U.; Helmstaedter, C.; Schramm, J.; Elger, C.E. Memory outcome after selective amygdalohippocampectomy in patients with temporal lobe epilepsy: One-year follow-up. *Epilepsia* **2004**, *45*, 960–962. [CrossRef] [PubMed]

63. Arts, S.; Delye, H.; van Lindert, E.J. Intraoperative and postoperative complications in the surgical treatment of craniosynostosis: Minimally invasive versus open surgical procedures. *J. Neurosurg. Pediatr.* **2017**, *24*, 1–7. [CrossRef] [PubMed]

64. Morcos, M.W.; Jiang, F.; McIntosh, G.; Johnson, M.; Christie, S.; Wai, E.; Ouellet, J.; Bailey, C.; Ahn, H.; Paquet, J.; et al. Predictors of Blood Transfusion in Posterior Lumbar Spinal Fusion: A Canadian Spine Outcome and Research Network Study. *Spine* **2018**, *43*, E35–E39. [CrossRef] [PubMed]

65. Duckworth, E.A.; Vale, F.L. Trephine epilepsy surgery: The inferior temporal gyrus approach. *Neurosurgery* **2008**, *63* (Suppl. 1), ONS156–ONS160. [CrossRef]

66. Schwab, R.S.; Sweet, W.H.; Mark, V.H.; Kjellberg, R.N.; Ervin, F.R. Treatment of intractable temporal lobe epilepsy by stereotactic amygdala lesions. *Trans. Am. Neurol. Assoc.* **1965**, *90*, 12–19. [PubMed]

67. Narabayashi, H.; Mizutani, T. Epileptic seizures and the stereotaxic amygdalotomy. *Confin. Neurol.* **1970**, *32*, 289–297. [CrossRef] [PubMed]

68. Flanigin, H.F.; Nashold, B.S. Stereotactic lesions of the amygdala and hippocampus in epilepsy. *Acta Neurochir.* **1976**, *23*, 235–239.

69. Heimberger, R.F.; Small, I.F.; Milstein, V.; Moore, D. Stereotactic amygdalotomy for convulsive and behavioural disorders. *Appl. Neurophysiol.* **1978**, *41*, 43–45.

70. Vaernet, K. Stereotaxic amygdalotomy in temporal lobe epilepsy. *Confin. Neurol.* **1972**, *34*, 176–180. [CrossRef] [PubMed]

71. Mempel, E.; Witkiewicz, B.; Stadnicki, R.; Luczywek, E.; Kuciński, L.; Pawłowski, G.; Nowak, J. The effect of medial amygdalotomy and anterior hippocampotomy on behavior and seizures in epileptic patients. *Acta Neurochir.* **1980**, *30*, 161–167.

72. Nadvornik, P.; Sramka, M.; Gajdosova, D.; Kokavec, M. Longitudinal hippocampectomy. *Confin. Neurol.* **1975**, *37*, 244–248. [CrossRef]

73. Parrent AG and Blume, W. Stereotactic Amygdalohippocampotomy for the Treatment of Medial Temporal Lobe Epilepsy. *Epilepsia* **1999**, *40*, 1408–1416. [CrossRef]

74. Le Bihan, D.; Delannoy, J.; Levin, R.L. Temperature mapping with MR imaging of molecular diffusion: Application to hyperthermia. *Radiology* **1989**, *171*, 853–857. [CrossRef] [PubMed]

75. Bertsch, F.; Mattner, J.; Stehling, M.K.; Müller-Lisse, U.; Peller, M.; Loeffler, R.; Weber, J.; Messmer, K.; Wilmanns, W.; Issels, R.; et al. Non-invasive temperature mapping using MRI: comparison of two methods based on based on chemical shift and T1-relaxation. *Magn. Reson. Imaging* **1998**, *16*, 393–404. [CrossRef]

76. Jermakowicz, W.J.; Kanner, A.M.; Sur, S.; Bermudez, C.; D'Haese, P.F.; Kolcun, J.P.G.; Cajigas, I.; Li, R.; Millan, C.; Ribot, R.; et al. Laser thermal ablation for mesiotemporal epilepsy: Analysis of ablation volumes and trajectories. *Epilepsia* **2017**, *58*, 801–810. [CrossRef] [PubMed]

77. Kang, J.Y.; Wu, C.; Tracy, J.; Lorenzo, M.; Evans, J.; Nei, M.; Skidmore, C.; Mintzer, S.; Sharan, A.D.; Sperling, M.R. Laser interstitial thermal therapy for medically intractable mesial temporal lobe epilepsy. *Epilepsia* **2016**, *57*, 325–334. [CrossRef] [PubMed]

78. Sagher, O.; Thawani, J.P.; Etame, A.B.; Gomez-Hassan, D.M. Seizure outcomes and mesial resection volumes following selective amygdalohippocampectomy and temporal lobectomy. *Neurosurg. Focus* **2012**, *32*, E8. [CrossRef] [PubMed]

79. Schramm, J. Temporal lobe epilepsy surgery and the quest for optimal extent of resection: A review. *Epilepsia* **2008**, *49*, 1296–1307. [CrossRef] [PubMed]

80. Renowden, S.A.; Matkovic, Z.; Adams, C.B.; Carpenter, K.; Oxbury, S.; Molyneux, A.J.; Anslow, P.; Oxbury, J. Selective amygdalohippocampectomy for hippocampal sclerosis: Postoperative MR appearance. *AJNR Am. J. Neuroradiol.* **1995**, *16*, 1855–1861. [PubMed]

81. Arruda, F.; Cendes, F.; Andermann, F.; Dubeau, F.; Villemure, J.G.; Jones-Gotman, M.; Poulin, N.; Arnold, D.L.; Olivier, A. Mesial atrophy and outcome after amygdalohippocampectomy or temporal lobe removal. *Ann. Neurol.* **1996**, *40*, 446–450. [CrossRef] [PubMed]

82. Pauli, E.; Pickel, S.; Schulemann, H.; Buchfelder, M.; Stefan, H. Neuropsychologic findings depending on the type of the resection in temporal lobe epilepsy. *Adv. Neurol.* **1999**, *81*, 371–377. [PubMed]

83. Clusmann, H.; Schramm, J.; Kral, T.; Helmstaedter, C.; Ostertun, B.; Fimmers, R.; Haun, D.; Elger, C.E. Prognostic factors and outcome after different types of resection for temporal lobe epilepsy. *J. Neurosurg.* **2002**, *97*, 1131–1141. [CrossRef] [PubMed]

84. Paglioli, E.; Palmini, A.; Portuguez, M.; Azambuja, N.; da Costa, J.C.; da Silva Filho, H.F.; Martinez, J.V.; Hoeffel, J.R. Seizure and memory outcome following temporal lobe surgery: Selective compared with nonselective approaches for hippocampal sclerosis. *J. Neurosurg.* **2006**, *104*, 70–78. [CrossRef] [PubMed]

85. Tanriverdi, T.; Olivier, A. Cognitive changes after unilateral cortico-amygdalohippocampectomy unilateral selective-amygdalohippocampectomy mesial temporal lobe epilepsy. *Turk. Neurosurg.* **2007**, *17*, 91–99. [PubMed]

86. Bate, H.; Eldridge, P.; Varma, T.; Wieshmann, U.C. The seizure outcome after amygdalohippocampectomy and temporal lobectomy. *Eur. J. Neurol.* **2007**, *14*, 90–94. [CrossRef] [PubMed]

87. Tanriverdi, T.; Dudley, R.W.; Hasan, A.; Al Jishi, A.; Al Hinai, Q.; Poulin, N.; Colnat-Coulbois, S.; Olivier, A. Memory outcome after temporal lobe epilepsy surgery: Corticoamygdalohippocampectomy versus selective amygdalohippocampectomy. *J. Neurosurg.* **2010**, *113*, 1164–1175. [CrossRef] [PubMed]

88. Wendling, A.S.; Hirsch, E.; Wisniewski, I.; Davanture, C.; Ofer, I.; Zentner, J.; Bilic, S.; Scholly, J.; Staack, A.M.; Valenti, M.P.; et al. Selective amygdalohippocampectomy versus standard temporal lobectomy in patients with mesial temporal lobe epilepsy and unilateral hippocampal sclerosis. *Epilepsy Res.* **2013**, *104*, 94–104. [CrossRef] [PubMed]

89. Bujarski, K.A.; Hirashima, F.; Roberts, D.W.; Jobst, B.C.; Gilbert, K.L.; Roth, R.M.; Flashman, L.A.; McDonald, B.C.; Saykin, A.J.; Scott, R.C.; et al. Long-term seizure, cognitive, and psychiatric outcome following trans-middle temporal gyrus amygdalohippocampectomy and standard temporal lobectomy. *J. Neurosurg.* **2013**, *119*, 16–23. [CrossRef] [PubMed]

90. Josephson, C.B.; Dykeman, J.; Fiest, K.M.; Liu, X.; Sadler, R.M.; Jette, N.; Wiebe, S. Systematic review and meta-analysis of standard vs selective temporal lobe epilepsy surgery. *Neurology* **2013**, *80*, 1669–1676. [CrossRef] [PubMed]

91. Hu, W.H.; Zhang, C.; Zhang, K.; Meng, F.G.; Chen, N.; Zhang, J.G. Selective amygdalohippocampectomy versus anterior temporal lobectomy in the management of mesial temporal lobe epilepsy: A meta-analysis of comparative studies. *J. Neurosurg.* **2013**, *119*, 1089–1097. [CrossRef] [PubMed]

92. Nascimento, F.A.; Gatto, L.A.; Silvado, C.; Mäder-Joaquim, M.J.; Moro, M.S.; Araujo, J.C. Anterior temporal lobectomy versus selective amygdalohippocampectomy in patients with mesial temporal lobe epilepsy. *Arq. Neuropsiquiatr.* **2016**, *74*, 35–43. [CrossRef] [PubMed]

93. Schmeiser, B.; Wagner, K.; Schulze-Bonhage, A.; Mader, I.; Wendling, A.S.; Steinhoff, B.J.; Prinz, M.; Scheiwe, C.; Weyerbrock, A.; Zentner, J. Surgical Treatment of Mesiotemporal Lobe Epilepsy: Which Approach is Favorable? *Neurosurgery* **2017**, *81*, 992–1004. [CrossRef] [PubMed]

94. Foged, M.T.; Vinter, K.; Stauning, L.; Kjær, T.W.; Ozenne, B.; Beniczky, S.; Paulson, O.B.; Madsen, F.F.; Pinborg, L.H.; Danish Epilepsy Surgery Group. Verbal learning and memory outcome in selective amygdalohippocampectomy versus temporal lobe resection in patients with hippocampal sclerosis. *Epilepsy Behav.* **2018**, *79*, 180–187. [CrossRef] [PubMed]

95. Goldtein, L.H.; Polkey, C.E. Behavioural memory after temporal lobectomy or amygdalo-hippocampectomy. *Br. J. Clin. Psychol.* **1992**, *31 Pt 1*, 75–81. [CrossRef]

96. Golstein, L.H.; Polkey, C.E. Short-term cognitive changes after unilateral temporal lobectomy or unilateral amygdalo-hippocampectomy for the relief of temporal lobe epilepsy. *J. Neurol. Neurosurg. Psychiatry* **1993**, *56*, 135–140. [CrossRef]

97. Wolf, R.L.; Ivnik, R.J.; Hirschorn, K.A.; Sharbrough, F.W.; Cascino, G.D.; Marsh, W.R. Neurocognitive efficiency following left temporal lobectomy: Standard versus limited resection. *J. Neurosurg.* **1993**, *79*, 76–83. [CrossRef] [PubMed]

98. Helmstaedter, C.; Elger, C.E. Cognitive consequences of two-thirds anterior temporal lobectomy on verbal memory in 144 patients: A three-month follow-up study. *Epilepsia* **1996**, *37*, 171–180. [CrossRef] [PubMed]

99. Hemstaedter, C.; Reuber, M.; Elger, C.C. Interaction of cognitive aging and memory deficits related to epilepsy surgery. *Ann. Neurol.* **2002**, *52*, 89–94. [CrossRef] [PubMed]

100. Hader, W.J.; Pillay, N.; Myles, S.T.; Partlo, L.; Wiebe, S. The benefit of selective over standard surgical resections in the treatment of intractable temporal lobe epilepsy. *Epilepsia* **2005**, *46*, 253–260.

101. Morino, M.; Uda, T.; Naito, K.; Yoshimura, M.; Ishibashi, K.; Goto, T.; Ohata, K.; Hara, M. Comparison of neuropsychological outcomes after selective amygdalohippocampectomy versus anterior temporal lobectomy. *Epilepsy Behav.* **2006**, *9*, 95–100. [CrossRef] [PubMed]

102. Helmstaedter, C.; Richter, S.; Roske, S.; Oltmanns, F.; Schramm, J.; Lehmann, T.N. Differential effects of temporal pole resection with amygdalohippocampectomy versus selective amygdalohippocampectomy on material-specific memory in patients with mesial temporal lobe epilepsy. *Epilepsia* **2008**, *49*, 88–97. [CrossRef] [PubMed]

103. Boucher, O.; Dagenais, E.; Bouthillier, A.; Nguyen, D.K.; Rouleau, I. Different effects of anterior temporal lobectomy and selective amygdalohippocampectomy on verbal memory performance of patients with epilepsy. *Epilepsy Behav.* **2015**, *52*, 230–235. [CrossRef] [PubMed]
104. Gül, G.; Yandim Kuşcu, D.; Özerden, M.; Kandemir, M.; Eren, F.; Tuğcu, B.; Keskinkiliç, C.; Kayrak, N.; Kirbaş, D. Cognitive Outcome after Surgery in Patients with Mesial Temporal Lobe Epilepsy. *Noro. Psikiyatr. Ars.* **2017**, *54*, 43–48. [CrossRef] [PubMed]
105. Yang, P.F.; Zhang, H.J.; Pei, J.S.; Lin, Q.; Mei, Z.; Chen, Z.Q.; Jia, Y.Z.; Zhong, Z.H.; Zheng, Z.Y. Keyhole epilepsy surgery: Corticoamygdalohippocampectomy for mesial temporal sclerosis. *Neurosurg. Rev.* **2016**, *39*, 99–108. [CrossRef] [PubMed]

brain sciences

MDPI

Case Report

Epilepsy Surgery for Skull-Base Temporal Lobe Encephaloceles: Should We Spare the Hippocampus from Resection?

Firas Bannout [1,*], Sheri Harder [2], Michael Lee [3], Alexander Zouros [4], Ravi Raghavan [5], Travis Fogel [6], Kenneth De Los Reyes [4] and Travis Losey [1]

[1] Department of Neurology, Loma Linda University Health, Loma Linda, CA 92354, USA; TLosey@llu.edu
[2] Department of Radiology (Division of Neuroradiology); Loma Linda University Health, Loma Linda, CA 92354, USA; SHarder@llu.edu
[3] Department of Radiology, Los Angeles County and University of Southern California Medical Center, Los Angeles, CA 90033, USA; mjl_202@med.usc.edu
[4] Department of Neurosurgery, Loma Linda University Health, Loma Linda, CA 92354, USA; AZouros@llu.edu (A.Z.); KDelosreyes@llu.edu (K.D.L.R.)
[5] Department of Pathology, Human Anatomy & Neurosurgery, Loma Linda University Health, Loma Linda, CA 92354, USA; RRaghavan@llu.edu
[6] Loma Linda Physical Medicine and Rehabilitation, Neuropsychology; Loma Linda University Health, Loma Linda, CA 92354, USA; TFogel@llu.edu
* Correspondence: fbannout@llu.edu

Received: 31 January 2018; Accepted: 8 March 2018; Published: 12 March 2018

Abstract: The neurosurgical treatment of skull base temporal encephalocele for patients with epilepsy is variable. We describe two adult cases of temporal lobe epilepsy (TLE) with spheno-temporal encephalocele, currently seizure-free for more than two years after anterior temporal lobectomy (ATL) and lesionectomy sparing the hippocampus without long-term intracranial electroencephalogram (EEG) monitoring. Encephaloceles were detected by magnetic resonance imaging (MRI) and confirmed by maxillofacial head computed tomography (CT) scans. Seizures were captured by scalp video-EEG recording. One case underwent intraoperative electrocorticography (ECoG) with pathology demonstrating neuronal heterotopia. We propose that in some patients with skull base temporal encephaloceles, minimal surgical resection of herniated and adjacent temporal cortex (lesionectomy) is sufficient to render seizure freedom. In future cases, where an associated malformation of cortical development is suspected, newer techniques such as minimally invasive EEG monitoring with stereotactic-depth EEG electrodes should be considered to tailor the surrounding margins of the resected epileptogenic zone.

Keywords: temporal lobe epilepsy; encephalocele; meningoencephalocele; tailored surgery

---

1. Introduction

An encephalocele refers to the protrusion of brain tissue, meninges, and cerebral spinal fluid (CSF) through a calvarial or skull base defect. Most of these lesions are due to congenital defects, although they can be acquired following trauma, surgery, or infection [1,2]. When associated with temporal lobe seizures, encephaloceles are associated with bony defects in the middle cranial fossa [3,4]. Associated defects have also been localized to the petrosal bone [5,6], cribriform plate [7], and diffusely throughout the skull.

Imaging is often reported to be normal, with the skull-base lesion noted only upon reassessment of initial neuroimaging [3,5,6,8,9]. Magnetic resonance imaging (MRI) and high-resolution CT scan imaging of the skull base allow for identification of these rare lesions and provide critical information needed to decide if a surgical approach is required as part of the treatment plan. High-resolution CT imaging has been shown to detect bony defects associated with temporal lobe encephaloceles that are not evident on MRI [3,8,10].

Patients with skull-base temporal encephaloceles may rarely present with seizures [3,4]. These seizures are usually due to involvement of the neocortex, and are frequently refractory to medical therapy [4,10,11]. Suggested mechanisms include tissue traction, gliosis, or associated dysplasia [7]. Associated cranial abnormalities are reported in only 15% of cases. These include band heterotopia, nodular heterotopia, diffuse cortical dysplasia, and schizencephaly [7].

Resection of the encephalocele with temporal neocortical excision (lesionectomy) or performing an anterior temporal lobectomy with amygdalohippocampectomy (ATLAH) has been reported to be effective in controlling seizures [3,4,7,8,10,12–14]. For example, in 18 cases collected by Faulkner et al., 2010, 56% underwent local excision and 44% underwent a wider excision with a lobectomy. The majority of extra temporal cases (33%) are treated with local excision, while the majority of temporal lobe cases (67%) were treated with lobectomy. There was no difference in the reported seizure freedom rates postoperatively (100% seizure freedom in all patients).

We reviewed 44 patients with epilepsy and surgically treated temporal encephalocele as described in the English-language medical literature (Table 1). Only one case report of bitemporal encephalocele was found, and the remaining cases were evenly divided between the left (54%) and the right (52%) temporal side. Sixty-three percent of cases were treated conservatively with lesionectomy or anterior temporal lobectomy (unclear extension of resected tissue), and 34% of cases were treated with ATLAH. The largest case report series were compiled by Panov et al., 2016 (six patients) [14] and Saavalainen et al., 2015 (twelve patients) [15].

Table 1. Reported surgically treated temporal lobe encephaloceles in patients with epilepsy.

Authors, Year	Age, Sex	Location	Congenital or Acquired	Surgical Approach	Pathology	Seizure Outcome/Duration
Ruiz Garcia et al., 1971 [9]	30, F	L temporal	Congenital	ATL	Gliosis and fibrosis	Free/NA
Hyson et al., 1984 [5]	40, W	R temporal	Acquired (after right mastoidectomy)	Lesionectomy	NA	Free/NA
	12, M	R temporal	Acquired (after trauma)	ATL	Mixed astrocytoma grade I and oligodendroglioma.	NA
	37, F	L temporal	Congenital	ATL	Gliosis	Free/3 y
Rosenbaum et al., 1985 [17]	38, F	R temporal	Congenital	ATL + AH	NA	Only Aura
	30, F	R temporal	Congenital	ATL + AH	NA	Free/NA
Whiting et al., 1990 [18]	18, F	R temporal	Congenital	ATL	Meningo-angiomatosis.	Only Aura
Le Blanc et al., 1991 [13]	37, F	L temporal	Congenital	ATL + AH	Gliosis	Free/NA
	36, M	L temporal	Congenital	ATL + AH	Gliosis	Free/NA
	26, M	L temporal	Acquired	ATL + AH	Gliosis	Free/NA
Wilkins et al., 1993 [19]	36, F	R temporal	Congenital	ATL	Gliosis	Free/18 Mo
Mulcahy et al., 1997 [20]	25, F	L temporal	Congenital	Lesionectomy	NA	NA
Yang et al., 2004 [6]	46, M	Bitemporal	Congenital	Lesionecotmy	Inflamed neuroglial	Free/NA
Byrne et al., 2010 [21]	26, M	L temporal	Congenital	Lesionectomy	Astrocystotis	Free/7 y
	42, F	L temporal	Congenital	ATL + AH	Astrocystotis	Free/2 y
	57, M	R temporal	Congenital	ATL + AH	Astrocystotis	Free/1 y
Aquilina et al., 2010 [12]	14, F	L temporal	Congenital (14)	ATL + AH	Diffuse temporal gliosis involving the HC + microdysgenesis of the amygdala.	Free
Abou-Hamden et al., 2010 [3]	39, F	L temporal	Congenital	ATL	Gliosis	Free/22 Mo
	26, F	L temporal	Congenital	ATL	Gliosis	Free/12 Mo
	26, F	L temporal	Acquired (forceps delivery)	ATL	Gliosis	Free/12 Mo

Table 1. *Cont.*

Authors, Year	Age, Sex	Location	Congenital or Acquired	Surgical Approach	Pathology	Seizure Outcome/Duration
Giulioni et al., 2014 [10]	41, M	L temporal	Congenital	ATL	Microdysgenesis	Free/5 y
	63, M	L temporal	Congenital	ATL	Microdysgenesis	Free/4 y
Gasparinin et al., 2014 [11]	20, M	L temporal	Congenital	Lesionectomy	Mild gliosis	Free/20 Mo
Shimada et al., 2015 [4]	21, M	L temporal	Congenital	Temporopolar disconnection	NA	Free/5 y
	36, M	L temporal	Congenital	Temporopolar disconnection	NA	Free/15 Mo
Saavalainen et al., 2015 [15]	22, M	L temporal	NA	ATL + disconnection	Gliosis	Free/2.5 y
	43, F	L temporal	NA	ATL + disconnection	Gliosis	Free/2.1 y
	45, F	R temporal	NA	ATL + disconnection	Gliosis	Engl II/0.79 y
	45, F	L temporal	NA	ATL + disconnection	Gliosis	Engle II/1.2 y
	32, M	L temporal	NA	ATL + disconnection	Gliosis	Engle 3 A/1.3 y
	40, M	R temporal	NA	ATL + disconnection	Gliosis	Free/1.1 y
	30, M	L temporal	NA	ATL + disconnection	Gliosis	Free/3 Mo
Saavalainen et al., 2015 [15] *	33, M	R temporal	NA	ATL + AH	Gliosis	Aura only/4.9 y
	31, M	R temporal	NA	ATL + AH	Gliosis	Free/6.2 y
	43, F	R temporal	NA	ATL + AH	Gliosis	Free/5.9 y
	44, F	R temporal	NA	ATL + AH	Gliosis	Free/3.2 y
	43, M	L temporal	NA	ATL + AH	Gliosis	Free/3.8 y
Panov et al., 2016 [14]	24, F	L temporal	NA	Disconnection (concern for verbal memory)	Mild gliosis	Free
	50, M	R temporal	NA	Lesionectomy (patient preference)	Reactive astrogliosis	Free
	45, F	R temporal	NA	ATL + AH (Szs originated from HC)	Severe astrogliosis	Free
	23, M	R temporal	NA	ATL +AH (increased volume or amygdala)	Mild gliosis	Engle IIb (recurred Szs but overall improved)
	45, F	R temporal	NA	ATL +AH (increase volume and T2 MRI signal of HC)	Mild Chaslin's gliosis, with FCD-IC affecting HC	Free
	39, M	R temporal	NA	ATL +AH (No reported MRI or EEG early involvement of HC)	Moderate gliosis and mild astrogliosis of HC	Free
De Souza et al., 2018 [16]	18, F	L temporal	NA	ATL + amygdalectomy alone	Architectural disorganization suggestive of FCD	Free/1 y

AH: amygdalohippocampectomy, ATL: anterior temporal lobectomy, EEG; electroencephalography, F: female, FCD-IC: focal cortical dysplasia, with abnormal radial and tangential cortical amination according to Blumcke's classification, HC: hippocampus, L: left, M: male, Mo: month, MRI: magnetic resonance imaging, NA: not available, R: right, Szs: seizures, T2: Time 2, y: year. * Samples from the base of the encephalocele were obtained from all 12 surgically treated patients. All of these samples showed gliosis and five patients (42%) had mild cortical laminar disorganization. The temporal lobe samples showed gliosis in 11 patients and heterotopic neurons in four patients. Hippocampus and amygdala were resected in five patients; the samples revealed normal mesial structures ($n = 2$), gliosis ($n = 2$), or small focal neuronal loss in CA2 area interpreted as possible mild hippocampal degeneration ($n = 1$).

Surgical approaches vary depending on seizure type, origin of encephalocele (congenital vs. acquired), location (temporal vs. extra-temporal), the type of brain lesion seen on MRI (with associated signal abnormality vs. none), and concordance of studies (scalp EEG MRI, positron emission tomography (PET) and neuropsychological testing). Most recently, stereotactic-depth EEG electrode implantation was used in one case report of temporal lobectomy and amygdalectomy alone, sparing the hippocampus, with reported seizure freedom [16].

2. Case Series

We report two cases of temporal lobe encephalocele, one congenital and another traumatic, which were treated with local excision and reparation of the encephalocele. Both cases underwent temporal lobe neocortical resection, sparing the hippocampus. Both patients have remained with Engel class-I surgical outcome (42 months for case 1 and 23 months for case 2).

2.1. Case 1: 24-Year-Old Ambidextrous, Bilingual (English and German) Man with Seizures Since 22 Years of Age. Seizure Semiology: Ictal Expressive Aphasia, Right Forced Head and Eyes Deviation, Lip-Smacking, and Sensation of Fear

Preoperative MRI (Figure 1A,B) and maxillofacial CT scan (Figure 1C) showed a large temporo-sphenoidal encephalocele with a defect in the floor of the middle cranial fossa with herniation of gliotic temporal and frontal opercular tissue into the defect.

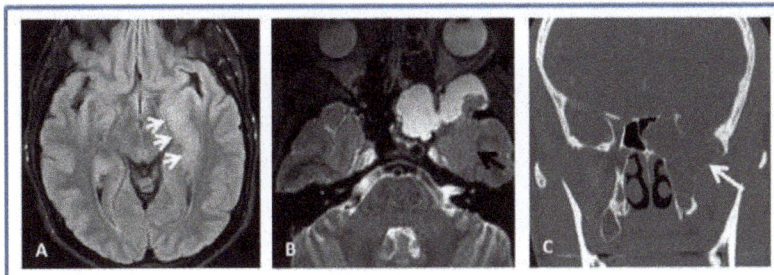

Figure 1. Preoperative brain imaging. (**A**) Axial FLAIR sequence showing increased signal in the left medial temporal lobe (small white arrows). (**B**) Axial T2 sequence shows herniated intracranial contents extending through the skull base defect. There is thickening of the regional cortex (black arrow). (**C**) Coronal CT maxillofacial showing the large left temporo-sphenoidal encephalocele (white arrows).

Scalp video-EEG monitoring captured five complex partial seizures, localized broadly to the left fronto-temporal channels (T3/T5/O1 and FP1/F3/F7). The ictal pattern was from EEG electrodecremental response, followed by 1.5–2 Hz rhythmic activity then 12–15 Hz low amplitude rhythmic activity (maximally at T5/O1), which is not typical for medial temporal lobe onset seizure but rather for neocortical onset seizure [22] (Figure 2). Electrocardiogram (ECG) correlated with tachycardia up to 120 bpm.

Intraoperative ECoG monitoring with three 2 × 4 grids were used to cover the left frontal operculum, anterior temporal tip containing the dysplastic tissue, and mid-posterior/lateral temporal cortex. One 1 × 8 strip was placed in the inferior medial and anterior temporal lobe. The strip and grid located in the inferior and anterior temporal lobes revealed occasional dysmorphic spikes anteriorly.

Epileptiform discharges were anteriorly located with the temporal pole and possibly the frontal opercular region. There was no evidence of epileptiform discharges coming from the medial-posterior temporal lobe strips.

Surgical approach included the insertion of a lumbar cerebrospinal fluid intrathecal drain, harvest of abdominal fat through a separate incision, left temporal craniotomy and temporal lobectomy with intraoperative electrocorticography, fronto-zygomatic craniotomy, and repair of left sphenonasal encephalocele with microscopic dissection. The ECoG findings allowed for identification of a 3-cm margin from the temporal tip for partial resection.

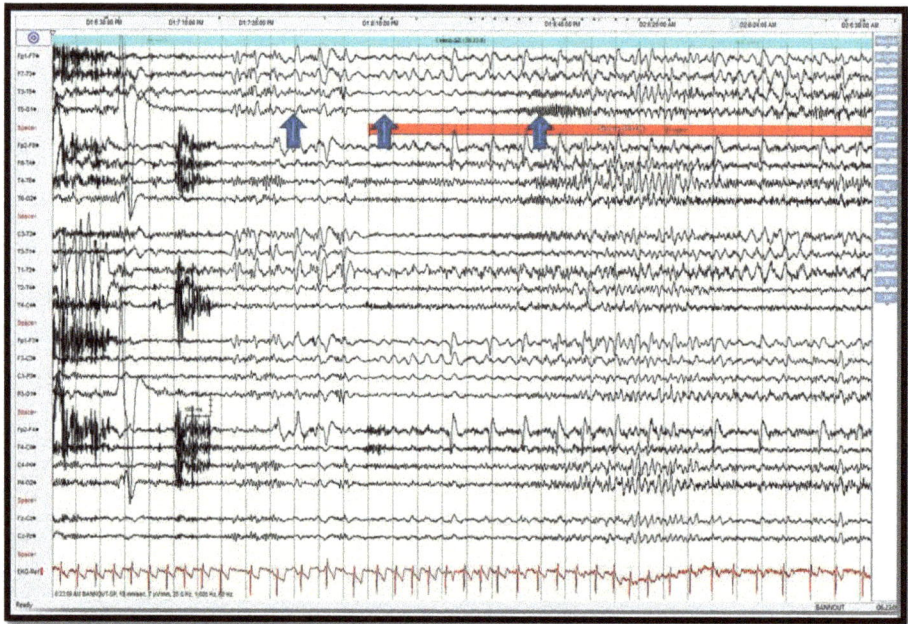

Figure 2. Scalp EEG recording showing broad periodic left fronto-temporal sharp waves (arrow 1 from the left) followed by attenuation at T3/T5 and T3/T1 (arrow 2) then low amplitude rhythmic beta frequencies at T5/O1 (arrow 3).

Surgical pathology of the resected tissue showed benign cerebral parenchyma and focal neuronal heterotropia within the deeper portion of the white matter (Figure 3). There was no evidence of gliosis or inflammation. There were no abnormal inclusions, myelin-related or metabolic abnormalities, or neoplastic changes.

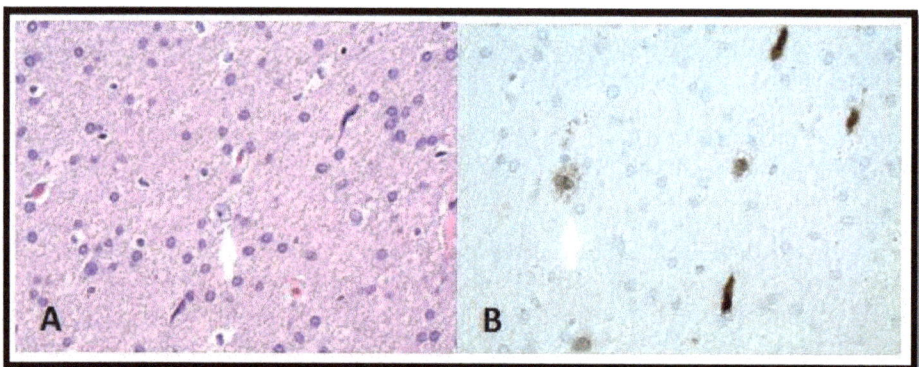

Figure 3. Histology of the resected left temporal lobe tissues. (**A**) Heterotopic white matter neurons HE stain × 400. (**B**) Heterotopic white matter neurons (5 in the field) highlighted by NeuN immunostain × 400.

Outcome: Seizure frequency prior to surgery was 4–5 times per week. He has remained seizure-free for 42 months since his surgery, during which levetiracetam was discontinued and lacosamide was reduced from 200 mg bid to 100 mg bid. Patient was able to continue his collegiate studies and scored 90th percentile on his *Medical College Admission Test* (MCAT). Post-operative MRI showed preserved left hippocampus (Figure 4A) and the left temporal surgical bed (Figure 4B). The amygdala was resected as well (Figure 4C).

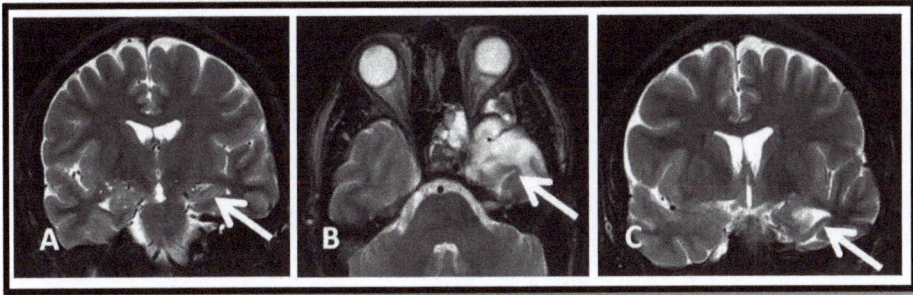

Figure 4. One year post operative brain MRI. (**A**) Coronal T2 sequence showing preserved left hippocampus (arrow). (**B**) Axial T2 sequence showing resected left temporal pole (arrow). (**C**) Coronal T2 sequence showing left amydalectomy (arrow).

2.2. Case 2: 66-Year-Old Right-Handed Man with Intractable Epilepsy since the Age of 43 Years, after He Was Struck on Forehead at Work by a Pump While Pumping Oil from a Compressor. Seizure Semiology Consisted of Dizziness, Starring Off, Lip-Smacking, Excessive Drooling, Right Hand Rolling Movements, and Occasional Bilateral Tonic-Clonic Seizures

Preoperative MRIs in 2009 and 2013 were reported as normal (1.5 Tesla). Preoperative high-resolution 3.0 Tesla MRI showed herniated right temporal pole encephalocele (Figure 5). Both hippocampi appeared normal and symmetric on MRI (Figure 5C).

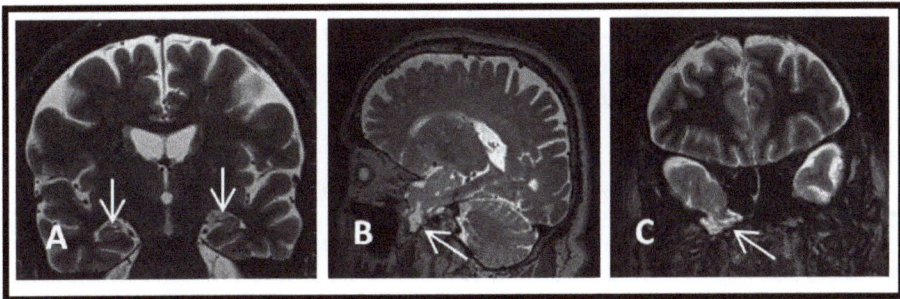

Figure 5. Preoperative brain MRI. (**A**) Coronal T2 sequence showing symmetrical bilateral hippocampi (vertical arrows). Coronal and Sagittal T2 sequence (**B**,**C**) showing small right sphenotemporal meningocele with small encephalocele and focal encephalomalacia of the inferior right temporal cortex (arrows).

Scalp video-EEG monitoring captured nine stereotyped electroclinical seizures (starring off, lip-smacking with occasional right manual automatisms). Ictal onset originated from the right anterior and mid-temporal channels. The electrographic ictal pattern consisted of rhythmic 2–3 Hz activity, suggesting neocortical rather than mesial temporal onset seizure pattern [22] (Figure 6).

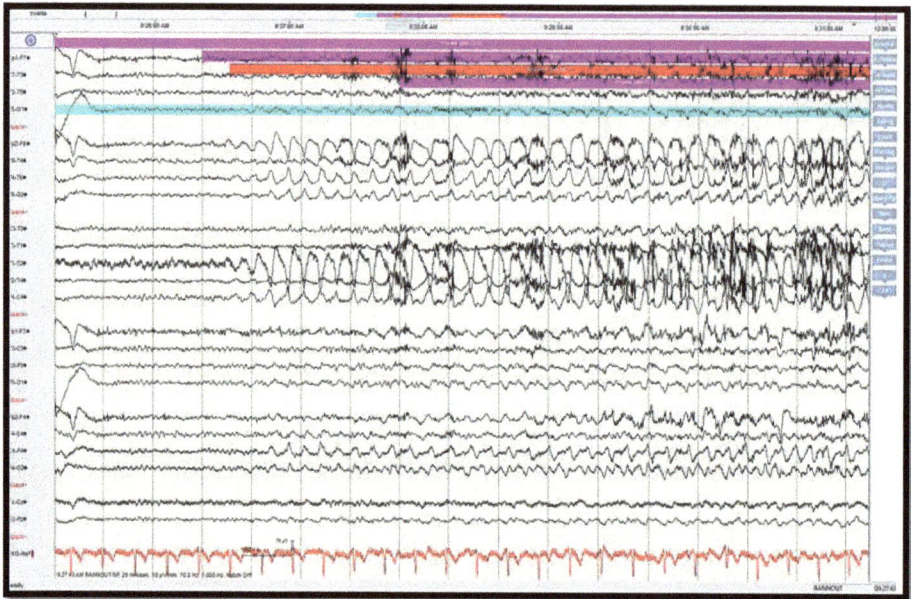

Figure 6. Scalp EEG recording showing right anterior temporal evolving 2–3 Hz ictal rhythm (arrows).

Neuropsychological testing showed relatively intact neuropsychological functioning with some areas of relative weaknesses, including evidence of weaker nonverbal (visual) new learning and memory compared to his auditory (verbal) new learning and memory as well as significant visuoconstructional weaknesses. The absence of any obvious nonverbal (visual) memory impairments provided support for the absence of any right mesial temporal dysfunction and the greater likelihood of seizure control with selective resection of the inferior temporal lobe.

Surgical approach included right temporal craniotomy, repair of lateral sphenoid encephalocele with repair of dural defect and rotational temporalis muscle and fascial flap, partial resection of right middle and inferior temporal gyrus for resection of seizure focus, and lumbar drain placement.

To address the seizure focus, the middle and inferior temporal gyri were identified and cauterized. Then, a corticotomy was performed followed by partial resection of the middle and inferior temporal gyri.

Outcome: Seizure occurred several times per month before surgery. He has remained seizure-free for 23 months since surgery. Seizure medications were reduced, during which lamotrigine was reduced from 900 mg to 600 mg daily and pregabalin was switched to gabapentin (from insurance perspective). Patient also has early onset Parkinson's disease and developed insomnia and restless leg syndrome. Clonazepam was added at 1 mg at night. Pathology was not done due to the use of cauterized resection. A post-operative MRI (Figure 7) showed the surgical cavity in the region of the previously noted encephalocele with a small residual fluid collection in the right infra-temporal fossa (Figure 7B).

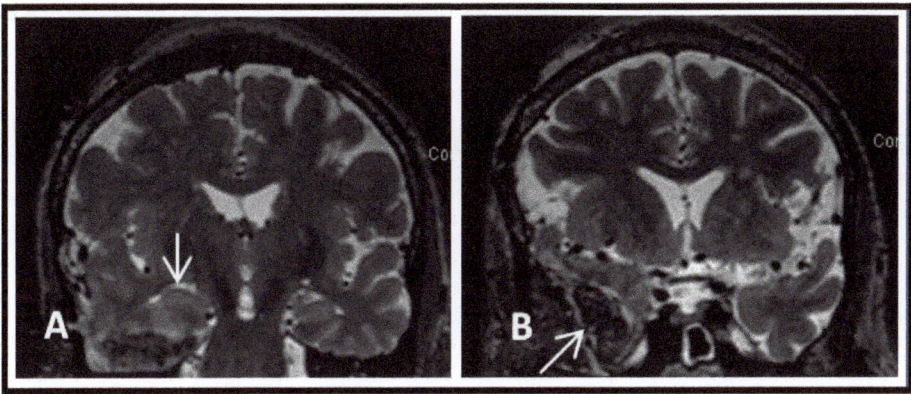

Figure 7. A post-operative brain MRI showing postsurgical changes related to resection of an encephalocele and the floor of the right middle cranial fossa. (**A**) Preserved right hippocampus (vertical arrow). (**B**) A surgical defect at the anterior aspect of the right temporal lobe (arrow).

3. Discussion

Meningoencephaloceles can be associated with microscopic congenital malformations. As such, it is important to realize that the meningoencephalocele may be the "the tip of the iceberg" and the diagnosis should lead one to carefully search for evidence of an underlying malformation of cortical development. However, seizure rates and associated cortical congenital abnormalities are probably more uncommon in this rare form of encephaloceles compared to non-skull base encephaloceles (cranial-vault) seen in pediatric patients, which commonly have multifocal cortical abnormalities and generalized epilepsy [23].

The optimal surgical strategy for intractable epilepsy patients with a temporal encephalocele is not well established. Tailored surgery of temporal pole encephaloceles was described by Giulioni et al. [10]. As it has been reported before, local excision (in extra-temporal cases) has a similar chance of seizure freedom as lobectomy (in temporal cases), regardless of the location of the encephalocele (Faulkner). In our review of nearly 44 cases published so far, a conservative surgical approach was seen more often than the larger resection of the anterior temporal lobe, including the amygdala and the hippocampus combined.

We describe here two cases of temporal lobe encephaloceles and related epilepsy in which tailored temporal lobe resection was performed. The left (dominant) hippocampus was preserved in case 1 due to the demanding high cognitive/memory function of a college student, and the right (non-dominant) hippocampus was preserved in case 2 due to patient choice and concern of worsening memory in the setting of his advanced age.

In the presence of electrographic or radiographic evidence of wide mesial structure involvement, the approach of ATLAH is reasonable, as shown by Panov et al., who reported four out of six surgical cases of temporal encephaloceles treated with ATLAH. However, two of those six cases were treated, similarly to our cases, with lesionectomy and temporal lobe disconnection. Our cases also illustrate that temporal lobe encephaloceles can be treated successfully with a conservative surgical approach sparing the mesial temporal structures.

The result of post-operative seizure freedom after encephalocele repair with only limited temporal lobe resection, sparing the hippocampus, demonstrates the possibility and the importance of preserving the vital memory function of the hippocampus without jeopardizing post-surgical seizure freedom.

As shown by De Souza et al., most recent advances in stereotactic depth electrode implantation may provide us with better options for intracranial EEG monitoring in future cases, which would allow us to identify the epileptogenic zone more precisely for better outcome

Acknowledgments: This case series was presented in a poster format at the American Epilepsy Society annual meeting in December 2015 and the Korean Congress of Radiology in May 2015.

Author Contributions: Bannout, F wrote the paper and was the first author and correspondent for this article. Harder, S and Lee, M provided the included radiographic images and participated in describing the nature and the radiographic appearance of temporal lobe encephaloceles. De Los Reyes, K and Zouros, a provided the neurosurgical techniques and description of the operational reports of two cases and additional comments on pediatric cranial vault encephaloceles. Raghavan, R provided the histopathologic images and comments of the operated two cases. Fogel, T described the neuropsycholgical analysis of case 2. Losey, T was the senior author who supervised writing the initial manuscript and analyzed the data and references thoroughly.

Conflicts of Interest: The authors declare no conflict of interest.

References

1. Jeevan, D.S.; Ormond, D.R.; Kim, A.H.; Meiteles, L.Z.; Stidham, K.R.; Linstrom, C.; Murali, R. Cerebrospinal fluid leaks and encephaloceles of temporal bone origin: Nuances to diagnosis and management. *World Neurosurg.* **2015**, *83*, 560–566. [CrossRef] [PubMed]
2. Mosnier, I.; Fiky, L.E.; Shahidi, A.; Sterkers, O. Brain herniation and chronic otitis media: Diagnosis and surgical management. *Clin. Otolaryngol. Allied Sci.* **2000**, *25*, 385–391. [CrossRef] [PubMed]
3. Abou-Hamden, A.; Lau, M.; Fabinyi, G.; Berkovic, S.F.; Jackson, G.D.; Mitchell, L.A.; Kalnins, R.; Fitt, G.; Archer, J.S. Small temporal pole encephaloceles: A treatable cause of "lesion negative" temporal lobe epilepsy. *Epilepsia* **2010**, *51*, 2199–2202. [CrossRef] [PubMed]
4. Shimada, S.; Kunii, N.; Kawai, K.; Usami, K.; Matsuo, T.; Uno, T.; Koizumi, T.; Saito, N. Spontaneous Temporal Pole Encephalocele Presenting with Epilepsy: Report of Two Cases. *World Neurosurg.* **2015**, *84*, e1–e6. [CrossRef] [PubMed]
5. Hyson, M.; Andermann, F.; Olivier, A.; Melanson, D. Occult encephaloceles and temporal lobe epilepsy: Developmental and acquired lesions in the middle fossa. *Neurology* **1984**, *34*, 363–366. [CrossRef] [PubMed]
6. Yang, E.; Yeo, S.B.; Tan, T.Y. Temporal lobe encephalocoele presenting with seizures and hearing loss. *Singap. Med. J.* **2004**, *45*, 40–42.
7. Faulkner, H.J.; Sandeman, D.R.; Love, S.; Likeman, M.J.; Nunez, D.A.; Lhatoo, S.D. Epilepsy surgery for refractory epilepsy due to encephalocele: A case report and review of the literature. *Epileptic Disord.* **2010**, *12*, 160–166. [PubMed]
8. Morone, P.J.; Sweeney, A.D.; Carlson, M.L.; Neimat, J.S.; Weaver, K.D.; Abou-Khalil, B.W.; Arain, A.M.; Singh, P.; Wanna, G.B. Temporal Lobe Encephaloceles: A Potentially Curable Cause of Seizures. *Otol. Neurotol.* **2015**, *36*, 1439–1442. [CrossRef] [PubMed]
9. Ruiz Garcia, F. A case of temporal lobe epilepsy caused by an encephalocele. *Rev. Esp. Otoneurooftalmol. Neurocir.* **1971**, *29*, 216–220. [PubMed]
10. Giulioni, M.; Licchetta, L.; Bisulli, F.; Rubboli, G.; Mostacci, B.; Marucci, G.; Martinoni, M.; Ferri, L.; Volpi, L.; Calbucci, F.; et al. Tailored surgery for drug-resistant epilepsy due to temporal pole encephalocele and microdysgenesis. *Seizure* **2014**, *23*, 164–166. [CrossRef] [PubMed]
11. Gasparini, S.; Ferlazzo, E.; Villani, F.; Didato, G.; Deleo, F.; Bellavia, M.A.; Cianci, V.; Latella, M.A.; Campello, M.; Giangaspero, F.; et al. Refractory epilepsy and encephalocele: Lesionectomy or tailored surgery? *Seizure* **2014**, *23*, 583–584. [CrossRef] [PubMed]
12. Aquilina, K.; Clarke, D.F.; Wheless, J.W.; Boop, F.A. Microencephaloceles: Another dual pathology of intractable temporal lobe epilepsy in childhood. *J. Neurosurg. Pediatr.* **2010**, *5*, 360–364. [CrossRef] [PubMed]
13. Leblanc, R.; Tampieri, D.; Robitaille, Y.; Olivier, A.; Andermann, F.; Sherwin, A. Developmental anterobasal temporal encephalocele and temporal lobe epilepsy. *J. Neurosurg.* **1991**, *74*, 933–999. [CrossRef] [PubMed]
14. Panov, F.; Li, Y.; Chang, E.; Knowlton, R.; Cornes, S. Epilepsy with temporal encephalocele: Characteristics of electrocorticography and surgical outcome. *Epilepsia* **2016**, *57*, e33–e38. [CrossRef] [PubMed]

Brain Sci. **2018**, *8*, 42

15. Saavalainen, T.; Jutila, L.; Mervaala, E.; Kälviäinen, R.; Vanninen, R.; Immonen, A. Temporal anteroinferior encephalocele: An underrecognized etiology of temporal lobe epilepsy? *Neurology* **2015**, *85*, 1467–1474. [CrossRef] [PubMed]

16. De Souza, J.P.S.A.S.; Mullin, J.; Wathen, C.; Bulacio, J.; Chauvel, P.; Jehi, L.; Gonzalez-Martinez, J. The usefulness of stereo-electroencephalography (SEEG) in the surgical management of focal epilepsy associated with "hidden" temporal pole encephalocele: A case report and literature review. *Neurosurg. Rev.* **2018**, *41*, 347–354. [CrossRef] [PubMed]

17. Rosenbaum, T.J.; Laxer, K.D.; Rafal, R.D.; Smith, W.B. Temporal lobe encephaloceles: Etiology of partial complex seizures? *Neurology* **1985**, *35*, 287–288. [CrossRef] [PubMed]

18. Whiting, D.M.; Awad, I.A.; Miles, J.; Chou, S.S.; Lüders, H. Intractable complex partial seizures associated with occult temporal lobe encephalocele and meningoangiomatosis: A case report. *Surg. Neurol.* **1990**, *34*, 318–322. [CrossRef]

19. Wilkins, R.H.; Radtke, R.A.; Burger, P.C. Spontaneous temporal encephalocele. Case report. *J. Neurosurg.* **1993**, *78*, 492–498. [CrossRef] [PubMed]

20. Mulcahy, M.M.; McMenomey, S.O.; Talbot, J.M.; Delashaw, J.B., Jr. Congenital encephalocele of the medial skull base. *Laryngoscope* **1997**, *107*, 910–914. [CrossRef] [PubMed]

21. Byrne, R.W.; Smith, A.P.; Roh, D.; Kanner, A. Occult middle fossa encephaloceles in patients with temporal lobe epilepsy. *World Neurosurg.* **2010**, *73*, 541–546. [CrossRef] [PubMed]

22. Ebersole, J.S.; Pacia, S.V. Localization of Temporal Lobe Foci by Ictal EEG Patterns. *Epilepsia* **1996**, *37*, 386–399. [CrossRef] [PubMed]

23. Bui, C.J.; Tubbs, R.S.; Shannon, C.N.; Acakpo-Satchivi, L.; Wellons, J.C., III; Blount, J.P.; Oakes, W.J. Institutional experience with cranial vault encephaloceles. *J. Neurosurg.* **2007**, *107* (Suppl. 1), 22–25. [PubMed]

brain sciences

MDPI

Article

L-Carnitine Modulates Epileptic Seizures in Pentylenetetrazole-Kindled Rats via Suppression of Apoptosis and Autophagy and Upregulation of Hsp70

Abdelaziz M. Hussein [1],[*] (ORCID), Mohamed Adel [1], Mohamed El-Mesery [2], Khaled M. Abbas [3], Amr N. Ali [3] and Osama A. Abulseoud [4]

[1]   Department of Medical Physiology, Faculty of Medicine, Mansoura University, Mansoura 35516, Egypt; madel7744@yahoo.com

[2]   Department of Biochemistry, Faculty of Pharmacy, Mansoura University, Mansoura 35516, Egypt; elmesery@hotmail.com

[3]   Faculty of Medicine, Mansoura University, Mansoura 35516, Egypt; kmakm_1@outlook.com (K.M.A.); amrnabmans@hotmail.com (A.N.A.)

[4]   Neuroimaging Research Branch, IRP, National Institute on Drug Abuse, National Institutes of Health, Biomedical Research Center, 251 Bayview Blvd, Baltimore, MD 21224, USA; osama.abulseoud@nih.gov

[*]   Correspondence: zizomenna28@yahoo.com or menhag@mans.edu.eg; Tel.: +20-1002421140

Received: 30 January 2018; Accepted: 9 March 2018; Published: 14 March 2018

Abstract: L-Carnitine is a unique nutritional supplement for athletes that has been recently studied as a potential treatment for certain neuropsychiatric disorders. However, its efficacy in seizure control has not been investigated. Sprague Dawley rats were randomly assigned to receive either saline (Sal) (negative control) or pentylenetetrazole (PTZ) 40 mg/kg i.p. × 3 times/week × 3 weeks. The PTZ group was further subdivided into two groups, the first received oral L-carnitine (L-Car) (100 mg/kg/day × 4 weeks) (PTZ + L-Car), while the second group received saline (PTZ + Sal). Daily identification and quantification of seizure scores, time to the first seizure and the duration of seizures were performed in each animal. Molecular oxidative markers were examined in the animal brains. L-Car treatment was associated with marked reduction in seizure score ($p = 0.0002$) that was indicated as early as Day 2 of treatment and continued throughout treatment duration. Furthermore, L-Car significantly prolonged the time to the first seizure ($p < 0.0001$) and shortened seizure duration ($p = 0.028$). In addition, L-Car administration for four weeks attenuated PTZ-induced increase in the level of oxidative stress marker malondialdehyde (MDA) ($p < 0.0001$) and reduced the activity of catalase enzyme ($p = 0.0006$) and increased antioxidant GSH activity ($p < 0.0001$). Moreover, L-Car significantly reduced PTZ-induced elevation in protein expression of caspase-3 ($p < 0.0001$) and β-catenin ($p < 0.0001$). Overall, our results suggest a potential therapeutic role of L-Car in seizure control and call for testing these preclinical results in a proof of concept pilot clinical study.

Keywords: L-carnitine; PTZ; epilepsy; apoptosis; β-catenin; oxidative stress

1. Introduction

Epilepsy is the second most common neurological issue after stroke affecting around 65 million people around the world [1,2]. As early as around 400 BC, it was proposed in the Hippocratic compositions that seizures start from the mind; however, the connection was not settled until seminal scientific investigations were conducted in the mid-Nineteenth Century by John Hughlings Jackson [3]. From that point forward, epileptology has grown significantly in parallel with fundamental discoveries in neuroscience.

Epilepsy is a neurological disease that is characterized by recurrent epileptic fits [4]. This disease involves several subtypes with particular phenotypes and pathophysiologies that are classified according to their etiology into symptomatic (secondary to other disorders, e.g., cancer or brain insult), idiopathic and cryptogenic (undetermined etiology, probably symptomatic) [5]. Basically, the term ictogenesis means the neurobiological basis for the transition of the brain from the interictal state to the ictal state, whereas epileptogenesis refers to the fundamental processes that lead to the development of the chronic phase of epilepsy with spontaneous recurrent seizures [6]. The molecular mechanisms behind ictogenesis and epileptogenesis are under intense investigation [7]. Current pharmacological treatment strategies largely aim at decreasing neuronal excitability and thereby preventing the occurrence of seizures. An alternative approach is to prevent the emergence of the epileptic state [8]. However, breakthrough seizures, treatment resistance [8], increased epilepsy-related morbidity and mortality [9] are fairly common, highlighting the fact that the efficacy of our current antiepileptogenic medications remains suboptimal [10]. This major public health concern calls for novel therapeutic targets for epilepsy [11].

L-carnitine (L-Car) is a dietary supplement that is available in health food stores. It is an endogenous molecule that plays an important role in β-oxidation via transport of acyl-moieties from fatty acids through the mitochondrial membrane [12]. After oral administration of L-Car, its plasma and cerebrospinal fluid (CSF) concentrations increase because it is easily transported through the blood-brain barrier via the organic cation/carnitine transporter OCTN2 [13]. Semland et al. [14] demonstrated that oral administration of L-Car in drinking water failed to stop the development of epilepsy in a kindling animal model, although it was associated with significant improvement of brain metabolism and energy production in pentylenetetrazole (PTZ)-induced epilepsy mouse model. Therefore, the present study was designed to explore the possible antiepileptic role of L-car and its effects on oxidative stress markers, apoptosis, autophagy and the neuroprotective heat shock proteins expression in the PTZ-kindled rat model.

2. Materials and Methods

2.1. Experimental Animals

Thirty male Sprague Dawely rats, 12–16-week-old at the beginning of the experiment, were housed in individual cages and had free access to food and water with a 12-h light-dark cycle. One week before the experiments, rats were adapted to these conditions with monitoring of the body weight and general conditions throughout the study. All protocols and experimental procedures were approved by the Institutional Review Board (IRB) at Mansoura Faculty of Medicine on 6 February 2017 with Approval Code # r/16.12.90.

2.2. Study Groups

Rats were randomly assigned to one of three groups ($n = 10$/group): 1, control negative: received saline (Sal group); 2, control positive pentylenetetrazole (PTZ); (Sal + PTZ group): rats received Sal and PTZ (40 mg/kg i.p. 3-times per week till full kindling) [15]; and 3, L-Car + PTZ group: same as the Sal + PTZ group, but rats received L-Car (100 mg/kg/day via gastric gavage) [16].

2.3. Pentylenetetrazole-Kindled Rat Model and Scoring of Epileptic Seizure

For induction of epilepsy, we used the method described by Hansen et al. [15]. After each PTZ injection, each rat was observed for 30 min for latency to epileptic fit, duration of seizure and seizure stage according to a modified Racine scale [15]. Full kindling was defined as exhibiting Stage 4 or 5 of seizure score on three consecutive trials.

2.4. Euthanasia and Collection of Brain Samples

Rats achieving full kindling were euthanized the following day after deep anesthesia using sodium thiopental (120 mg/kg i.p.). Five rats in each group were perfused transcardially with 100 mL heparinized saline followed by 150 mL of 10% formalin. The brain was then collected and placed in 10% paraformaldehyde for 4 h for fixation before it was stored in a 25% sucrose plus 0.1% sodium azide solution until processing. For the biochemical assay of markers of oxidative stress and Western blotting, whole brain tissues (of the remaining 5 rats in each group) were collected after saline perfusion only and stored in liquid nitrogen until the time of the required experiments.

2.5. Assay of Lipid Peroxidations Marker (MDA) and Antioxidants (GSH Activity) and Catalase in Brain Tissues

About 50–100 mg of brain tissues were homogenized using a mortar and pestle in 1–2 mL of cold buffer solution (50 mM potassium phosphate, pH 7.5, 1 mM ethylenediaminetetraacetic acid (EDTA)) then centrifuged at 4000 rpm for 15 min at 4 °C. The supernatant was kept at −20 °C until used for analysis. Malondialdehyde (MDA), reduced glutathione (GSH) and catalase enzyme activity in the supernatant of kidney homogenates were measured using a colorimetric method according to the manufacturer's instructions (Bio-Diagnostics, Dokki, Giza, Egypt).

2.6. Gel Electrophoresis and Western Blotting for Caspase-3 and β-Catenin

Brain tissues were lysed using RIPA buffer solution (50 mM Tris-HCl pH 7.4, 1% triton-X, 0.1% SDS, 150 mM NaCl, 2 mM EDTA, 50 mM NaF) containing protease inhibitor (Roche Diagnostics, Mannheim, Germany) and phosphatase inhibitor mixture II (Sigma Aldrich, Deisenhofen, Germany) to obtain total cell lysates. Afterwards, protein concentrations in each sample were assayed using the Bradford protein assay, and 30 μg of each sample were mixed with Laemmli buffer (pH 8.0) containing 8% sodium dodecyl sulfate (SDS), 0.2 M Tris, 10% β-mercaptoethanol and 40% glycerol. Then, protein samples were denatured by heating at 95 °C for 5 min. Vertical SDS-PAGE (polyacrylamide) electrophoresis was performed to fractionate total cell lysate proteins according to their molecular weight. Then, the fractionated proteins were transferred to the nitrocellulose membrane by the wet blotting method. Afterwards, membranes were blocked by incubation with 5% BSA solution for 1 h at room temperature. Samples were incubated at 4 °C overnight with primary antibodies specific for caspase-3 (molecular weight (MW) 35 kDa, #9662, Cell Signaling Technology, Danvers, MA, USA), β-catenin (MW 92 kDa, clone 6B3, Cell Signaling Technology, Danvers, MA, USA) and tubulin-α (MW 50 kDa, Neomarkers, Fremont, CA, USA). The secondary antibodies were anti-mouse-HRP (Dako-Cytomation, Glostrup, Denmark) and anti-rabbit-HRP (Cell Signaling Technology, Danvers, MA, USA). Finally, membranes were visualized using the ECL Western blotting detection system (Thermo Fisher Scientific, Waltham, MA, USA) according to the manufacturer's instructions.

2.7. Histopathological Examination of Hippocampal Neurons by Hematoxylin and Eosin

Twenty micrometer-thick sections of brain slices were stained with hematoxylin and eosin (H&E), and the slides of hippocampus were stained with hematoxylin for 15 min and in HCl alcohol solution for 35 s. Then, the sections were immersed with eosin for 10 min and 90% ethanol for 40 s. After that, the section was examined, and images of the cornu ammonis (CA3) region were captured under light microscope.

2.8. Measurement of Expression of Heat Shock Protein (Hsp) 70 and Microtubule-Associated Protein 1A/1B-Light Chain 3 (LC3) by Immunohistochemistry in Hippocampus

Serial coronal sections (40 μm) were sliced using a freezing sledge microtome, and a 1:6 series was used for all quantitative immunohistochemistry. Peroxidase-based immunostaining was completed as described previously [17]. In brief, following quenching of endogenous peroxidase activity where appropriate (using a solution of 3% hydrogen peroxide/10% methanol in distilled water) and blocking of nonspecific secondary antibody binding (using 3% normal serum in Tris-buffered saline (TBS) with 0.2% Triton X-100 at room temperature for 1 h), sections were incubated overnight at room temperature with the appropriate primary antibody diluted in 1% normal serum in TBS with 0.2% Triton X-100 (polyclonal anti-LC3 rabbit antibody, Cat#YPA1340, dilution 1:200; Chongqing Biospes Co., Ltd., Chongqing, China) and with primary rabbit polyclonal anti-Hsp70 (Cat #NBP1-35969, Novus Bio, Littleton, CO, USA) (diluted 1:50). Processing of brain sections and the process of immunostaining and image capture and analysis were mentioned in full detail in our previous work [18].

2.9. Statistical Analysis

Behavioral data (i.e., seizure scores, time to first seizure and the duration of seizures) are presented as the mean ± standard errors of mean (SEM). Molecular data on the effects of L-Car on oxidants and antioxidants are also presented as the mean ± SEM. Separate repeated measures analyses of variance (ANOVA) with treatment (saline and L-Car) and time (days) factors were used to compare each behavioral variable between control positive (PTZ + saline) and experimental (PTZ + L-Car) groups. When a significant interaction was detected, post-hoc *t*-tests were used to compare the two groups at different time points. Two separate survival analyses were used to examine differences in the time to first seizure and differences in the duration of seizure between the two groups. Separate one-way analyses of variance (ANOVA) were used to compare each molecular variable between control negative (saline), control positive (PTZ + saline) and experimental (PTZ + L-Car) groups. Pearson correlation analyses were used to study the relationships between seizure stage and individual oxidative stress markers in the PTZ group. All data were analyzed using GraphPad Prism Version 7 software, (GraphPad Software, La Jolla, CA, USA). Results were considered significant when $p \leq 0.05$.

3. Results

3.1. The Behavioral Effects of L-Car on PTZ-Induced Seizure

L-Car treatment was associated with marked reduction in seizure score ($F (1, 16) = 22.3, p = 0.0002$) that was evident as early as Day 2 of treatment (L-Car + PTZ vs. Sal + PTZ Day 1 mean ± SEM = 1.2 ± 0.4 vs. 2.4 ± 0.17 $t = 2.61$ df = 16, $p = 0.018$) and continued throughout treatment (Day 14: 1.4 ± 0.5 vs. 4.2 ± 0.2, $t = 4.64$ df = 16, $p = 0.0003$; Figure 1A). Furthermore, L-Car significantly prolonged the time to the first seizure (median survival time L-Car + PTZ vs. Sal + PTZ = 165 vs. 100 s, $X^2 = 31.07$, df = 1, $p < 0.0001$; Figure 1B) and shortened seizure duration (median survival time L-Car + PTZ vs. Sal + PTZ = 30 vs. 35 s, $X^2 = 4.81$, df = 1, $p = 0.028$; Figure 1C). Furthermore, we studied the relationship between seizure latency and seizure score and found a significant inverse correlation ($r = -4.96, p < 0.0001$, $n = 91$; Figure 1D). Moreover, animals with short latency (≤ 100 s) compared to long latency (>100 s) have a significantly higher seizure score ($t = 5.739$, df = 88, $p < 0.0001$; Figure 1E).

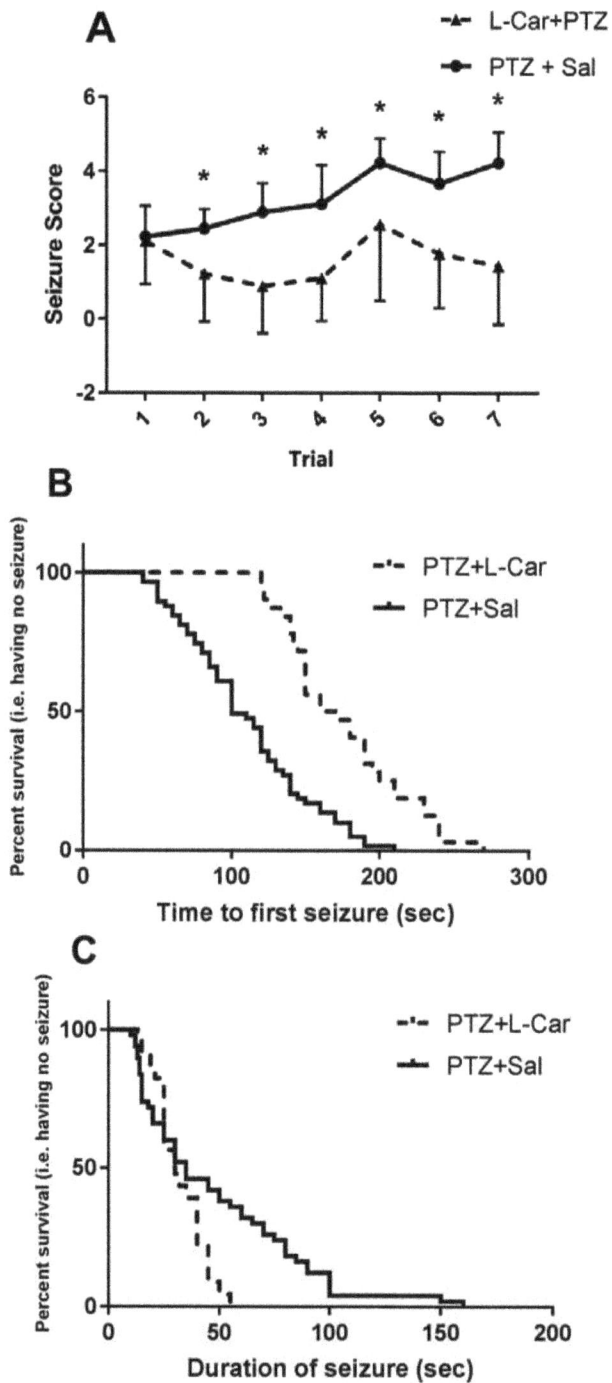

Figure 1. *Cont.*

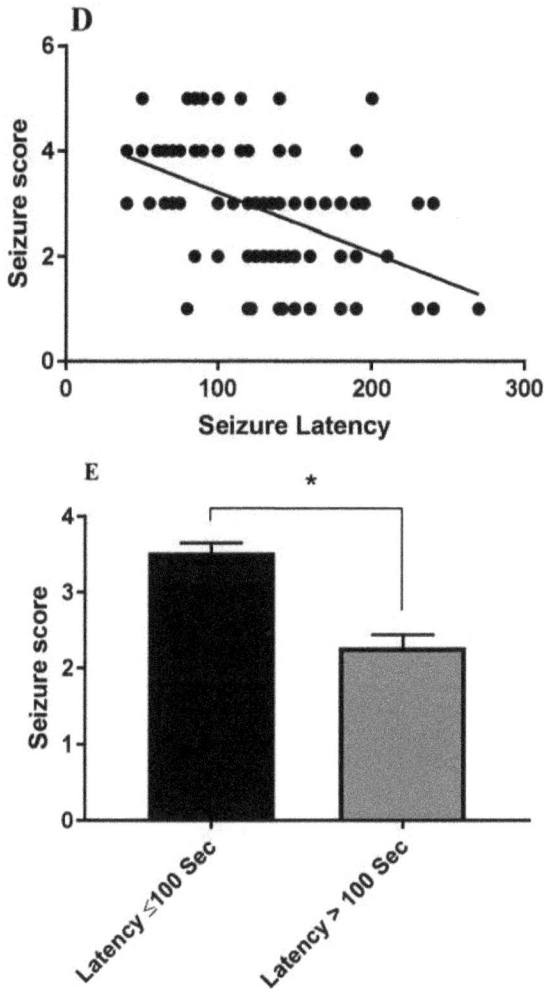

Figure 1. The behavioral effects of L-Car on PTZ-induced seizures. (**A**) Two-way ANOVA showed significant effects of treatment ($F_{1, 16}$ = 22.35, *p* = 0.0002), time ($F_{6, 96}$ = 6.07, *p* < 0.0001) and the interaction between the two factors ($F_{6, 96}$ = 3.15, *p* = 0.007) with significant differences (*) in seizure scores between L-Car + PTZ and Sal + PTZ groups on Day 2 (1.22 ± 0.43 vs. 2.4 ± 0.17, df = 16, *p* = 0.018), Day 3 (0.88 ± 0.42 vs. 2.8 ± 0.26, df = 16, *p* = 0.001), Day 4 (1.1 ± 0.38 vs. 3.1 ± 0.35, df = 16, *p* = 0.005), Day 5 (2.56 ± 0.6 vs. 4.2 ± 0.2, df = 16, *p* = 0.035), Day 6 (1.7 ± 0.49 vs. 3.6 ± 0.28, df = 16, *p* = 0.004) and Day 7 (1.4 ± 0.5 vs. 4.2 ± 0.2, *t* = 4.64, df = 16, *p* = 0.0003) by the two-tailed *t*-test, nine animals per group. (**B**) Survival analysis shows significant delay in time to first seizure in L-Car + PTZ vs. Sal + PTZ animals. Median survival ratio = 1.65 (165 vs. 100 s), 95% CI of ratio = 1.073–2.537, *p* < 0.0001. (**C**) Seizure duration is significantly reduced in L-Car + PTZ vs. Sal + PTZ animals. Median survival ratio is 0.85% (30 vs. 35 s), 95% CI of ratio = 0.523–0.404, *p* = 0.028. (**D**) Correlation between seizure latency and seizure score (*r* = −4.96, *p* < 0.0001, *n* = 91). (**E**) Seizure score in animals with short latency (≤100 s) and long latency (>100 s) (*t* = 5.739, df = 88, *p* < 0.0001). (*) significant in seizure scores between rats with sort latency and long latency. PTZ: pentylenetetrazole; Sal: saline; L-Car: L-carnitine.

3.2. The Molecular Effects of L-Car on MDA, GSH, Catalase Enzyme, Caspase-3 and β-Catenin

L-Car administration for 14 days attenuated PTZ-induced increase in MDA level ($F_{(2, 15)}$ = 98.51, $p < 0.0001$; Figure 2A), reduced the activity of catalase enzyme ($F_{(2, 15)}$ = 12.76, $p = 0.0006$; Figure 2B) and increased GSH concentration ($F_{(2, 15)}$ = 117.6, $p < 0.0001$; Figure 2C). Moreover, L-Car significantly reduced PTZ-induced elevation in protein expression of caspase-3 ($F_{(2, 15)}$ = 348.6, $p < 0.0001$; Figure 2D) and β-catenin ($F_{(2, 15)}$ = 1813, $p < 0.0001$; Figure 2E).

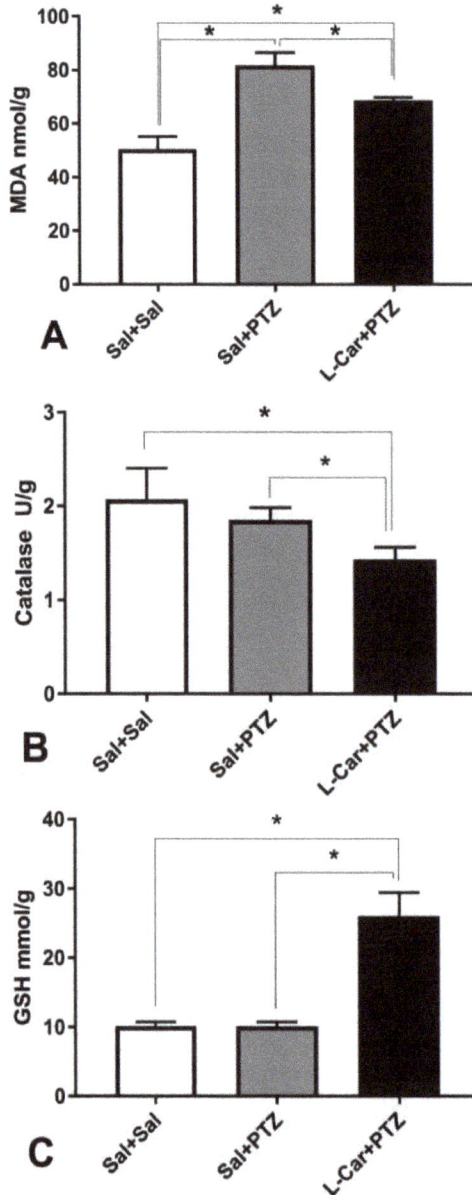

Figure 2. *Cont.*

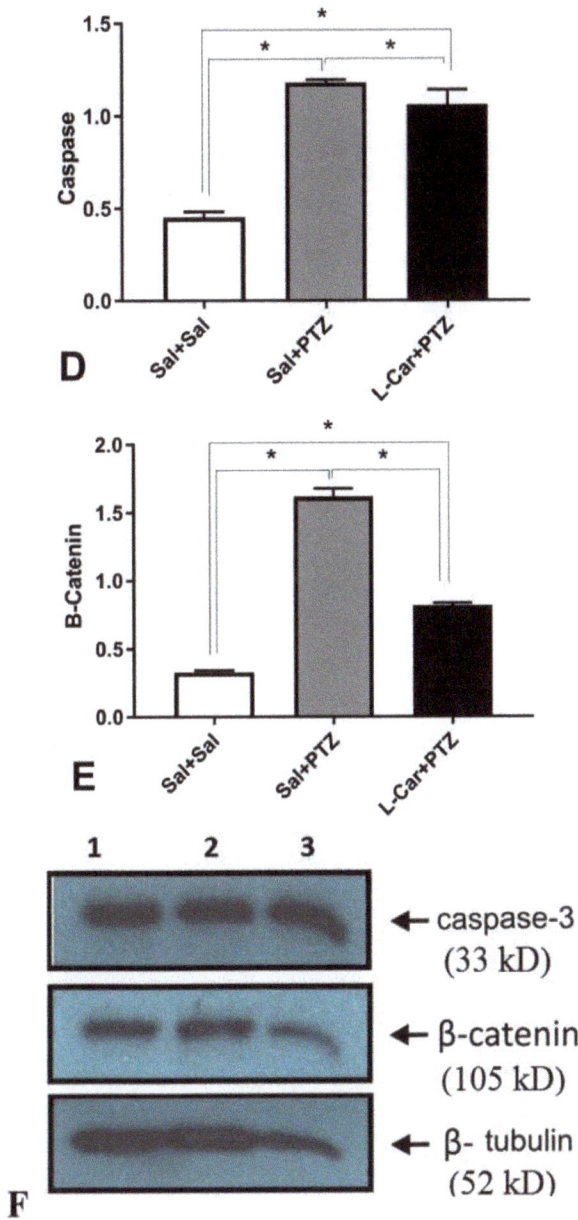

Figure 2. The molecular effects of L-car on oxidants and antioxidants. (**A**) PTZ-treated animals had significantly (*) higher malondialdehyde (MDA) compared to saline controls (*): mean difference = 31.35, 95% CI = 25.52–37.18, $p < 0.0001$, and L-Car-treated PTZ animals compared to saline controls (*): mean difference = 18.22, 95% CI = 12.39–24.04, $p < 0.0001$, as well as compared to PTZ-treated animals: mean difference = −13.13, 95% CI = −18.96–−7.307, $p < 0.0001$, by one-way ANOVA followed by Tukey's multiple comparisons test, six animals per group. (**B**) Significant difference in catalase enzyme activity ($F_{2, 15} = 12.76$, $p = 0.0006$) by one-way ANOVA with marked reduction in L-Car-treated animals

(compared to saline control (*): mean difference = −0.63, 95% CI = −0.97−−0.30, p = 0.0005, and compared to PTZ-treated animals (*): mean difference = −0.42, 95% CI = −0.75−−0.08, p = 0.013). No significant difference between PTZ-treated animals and saline controls (mean difference = −0.21, 95% CI = −0.55−0.11, p = ns) by Tukey's multiple comparisons test, six animals per group. (**C**) Robust increase in GSH concentration ($F_{2, 15}$ = 117.6, p < 0.0001) by one-way ANOVA. L-Car-treated animals showed a marked increase compared to saline-treated animals (*): mean difference = 15.8, 95% CI = 12.7–18.9, p < 0.0001, and compared to PTZ-treated animals (*): mean difference = 15.8, 95% CI = 12.7–18.9, p < 0.0001. No significant difference between PTZ-treated animals and saline controls (mean difference = 0, 95% CI = −0.31−3.1, p = ns) by Tukey's multiple comparisons test, six animals per group. (**D**) Significant difference between groups in caspase-3 protein expression ($F_{2, 15}$ = 348.6, p < 0.0001) by one-way ANOVA. PTZ increased the expression compared to saline controls (*): mean difference = 0.72, 95% CI = 0.65–0.80, p < 0.0001. Furthermore, L-Car increased the expression of caspase-3 protein compared to saline controls (*): mean difference = 0.60, 95% CI = 0.53–0.68, p < 0.0001. However, L-Car treatment attenuated the PTZ-induced increase in caspase-3 protein expression (*): L-Car + PTZ vs. Sal + PTZ mean difference = −0.12, 95% CI = −0.197−−0.043, p = 0.002, by Tukey's multiple comparisons test, six animals per group. (**E**) Marked effect of treatment on β-catenin ($F_{2, 15}$ = 1813, p < 0.0001) by one-way ANOVA. PTZ increased the expression of β-catenin compared to saline control (*): mean difference = 1.28, 95% CI = 1.23–1.34, p < 0.0001. Furthermore, L-Car increased the protein expression of β-catenin compared to saline controls (*): mean difference = 0.48, 95% CI = 0.43–0.54, p < 0.0001. However, L-Car treatment attenuated the PTZ-induced increase in β-catenin protein expression (*): L-Car + PTZ vs. Sal + PTZ mean difference = −0.80, 95% CI = −0.85−−0.74, p < 0.0001, by Tukey's multiple comparisons test, six animals per group, (**F**) Products of Western blot for caspase-3, β-catenin protein and tubulin (housekeeping gene) protein expression in the Sal group [1], Sal + PTZ group [2] and L-Car + PTZ group [3].

3.3. Effects of L-Car on the Morphology of Neurons in the CA3 Region of Hippocampus

Brain sections from the Sal group (Figure 3A) showed a normal appearance for neurons in the CA3 region of hippocampus, while those obtained from the Sal + PTZ group (Figure 3B) showed irregular arrangement of neurons with a reduction in the number of neurons, and the neurons showed pyknosis (darkly-stained nucleus and cytoplasm). The number of abnormal neurons was significantly reduced in the brains tissues obtained from the L-Car + PTZ group (Figure 3C).

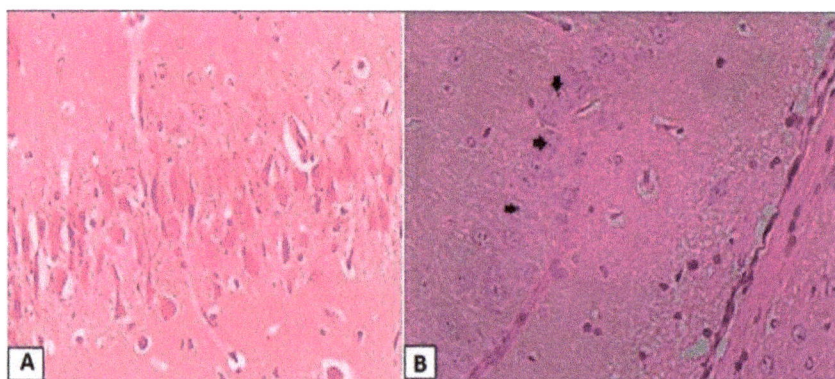

Figure 3. *Cont.*

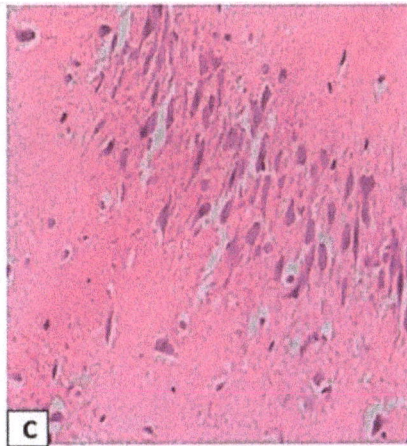

Figure 3. Histopathological examination of the CA3 region in hippocampus. Normal morphology for cells and adequate number in CA3 in brains obtained from the Sal group (**A**, H&E, 400×), reduced number of neurons with pyknotic changes (black arrows) in neurons of CA3 in brains obtained from the Sal + PTZ group (**B**, H&E, 400×) and significant number of neurons with normal morphology in CA3 in brains obtained from the L-Car + PTZ group (**C**, H&E, 400×).

3.4. Effects of L-Car on Hsp70 and LC3 Expression in the CA3 Region of Hippocampus

Immunohistochemical examination showed a significant decrease in the mean area of interest (AOI) of Hsp70-positive cells in the CA3 region of hippocampus in the Sal + PTZ and L-Car + PTZ groups compared to the Sal group ($p < 0.001$). On the other hand, the L-Car + PTZ group showed a significant increase in the number of Hsp70-positive cells compared to the Sal + PTZ group ($p < 0.01$) (Figure 4A). Brain sections showed a high expression of Hsp70 in the Sal group (Figure 4B), mild cytoplasmic expression in the Sal + PTZ group (Figure 4C) and high cytoplasmic expression in the L-Car + PTZ group (Figure 4D).

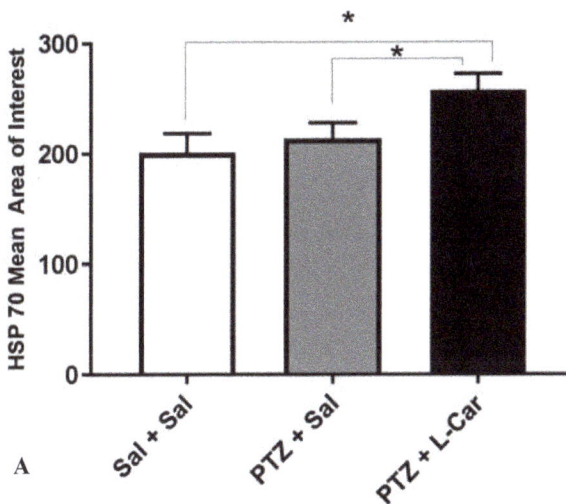

Figure 4. *Cont.*

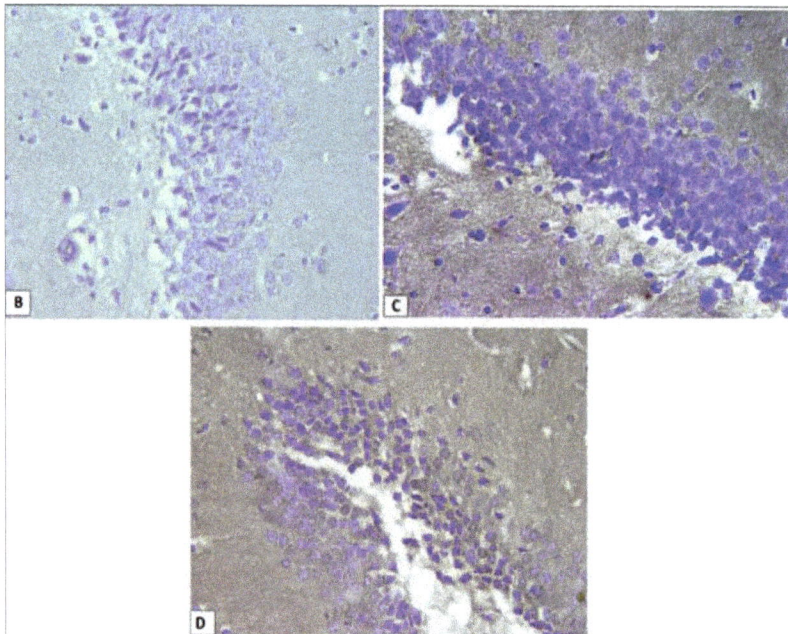

Figure 4. Mean area of interest of Hsp70 positivity in the CA3 region of hippocampus from different groups (**A**). The HSP70 protein expression mean area of interest was compared between the three groups (PTZ + Sal, Sal + Sal and PTZ + L-Car using one-way ANOVA. $F(2,15)=20.53$, $p < 0.0001$. Tukey multiple comparisons test was done between the groups. There was a significant difference between PTZ + Sal vs. PTZ + L-Car (*) mean difference = -43.67. 95% CI of difference = -67.47 to -19.86, $p = 0.0007$) and also between Sal + Sal vs. PTZ + L-Car (*) mean difference = -55.83. 95% CI of difference = -79.64 to -32.03, $p < 0.0001$. Section of brain in the CA3 region of hippocampus showing negative expression for Hsp70 in the Sal group (**B**, 200×), moderate cytoplasmic expression in the Sal + PTZ group (**C**, 200×) and high cytoplasmic brown staining for Hsp70 in the L-Car + PTZ group (**D**, 200×).

Furthermore, immunohistochemical examination showed a significant increase in the mean area of interest (AOI) of LC3-positive cells in the CA3 region of hippocampus in the Sal + PTZ group compared to the Sal group ($p < 0.001$). On the other hand, the L-Car + PTZ group showed a significant decrease in the LC3 compared to the Sal + PTZ group ($p < 0.01$) (Figure 5A). Brain sections showed negative expression of LC3 in the Sal group (Figure 5B), high cytoplasmic expression in the Sal + PTZ group (Figure 5C) and minimal cytoplasmic expression in the L-Car + PTZ group (Figure 5D).

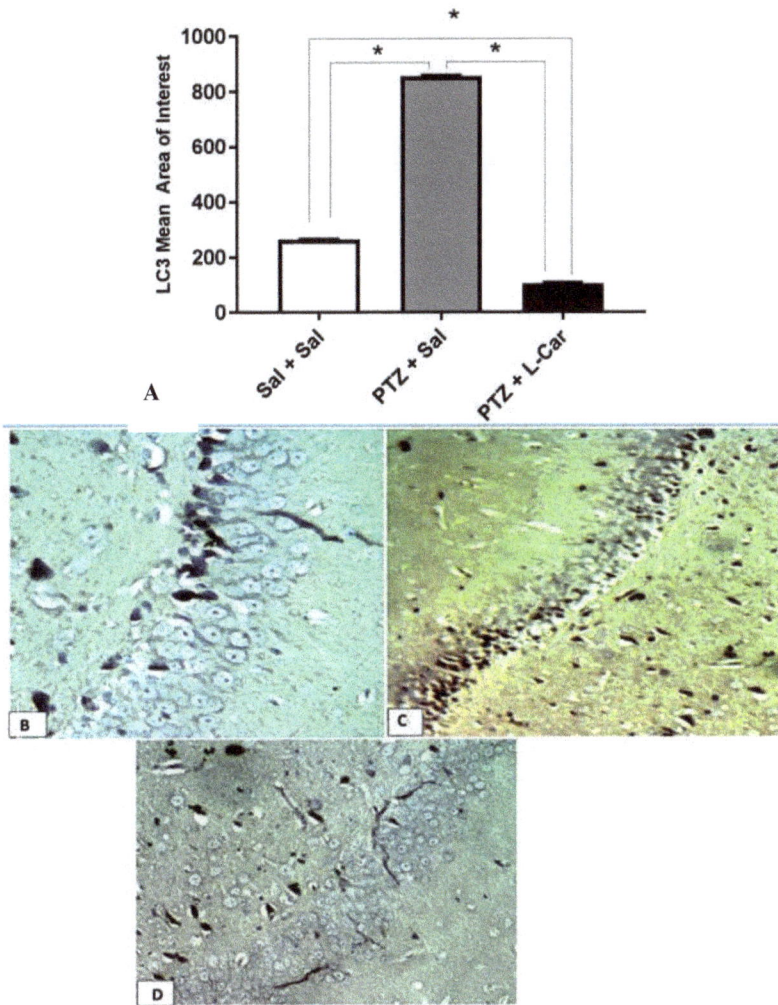

Figure 5. Mean area of interest of LC3 positivity in the CA3 region of hippocampus from different groups (**A**). LC mean area of interest by one-way ANOVA F(2,15) = 72,703, *p* < 0.0001. Tukey multiple comparisons test was done between the groups. There was a significant difference between PTZ + Sal vs. PTZ + L-Car (*) mean difference = 750.8. 95% CI of difference = 745.5 to 756.2, *p* < 0.0001 and also between Sal + Sal vs. PTZ + L-Car (*) mean difference = 160.4. 95% CI of difference = 155 to 165.8, *p* < 0.0001, and also between PTZ + Sal vs. Sal + Sal (mean difference = 590.4 95% CI of difference = 585 to 595.8, *p* < 0.0001). Section of brain in the CA3 region of hippocampus showing negative expression for LC3 in the Sal group (**B**, 400×), high cytoplasmic expression in the Sal + PTZ group (**C**, 400×) and low expression for LC3 in the L-Car + PTZ group (**D**, 400×).

3.5. Correlations between the Stage of Seizure, Oxidative Stress Markers, Hsp70 Expression and LC3 Expression in Hippocampus

Seizure stage showed significant positive correlations between seizure stage and MDA, LC3, caspase-3 and Hsp70 with significant negative correlations with GSH and CAT (*p* ≤ 0.01). Furthermore, expression of LC3 and Hsp70 showed a significant positive correlation with MDA and a significant negative correlation with CAT ($p \leq 0.01$). In addition, LC3 expression showed significant positive correlation with Hsp70 expression ($p < 0.01$) (Table 1).

Table 1. Correlation between seizure stage and oxidative stress markers, expression of caspase-3, β-catenin, LC3 and Hsp70 in the PTZ group.

Parameters	Stage of Seizure	MDA	GSH	CAT	Caspase-3	β-Catenin	Hsp70	LC3
Seizure stage	*r*	0.78	−0.74	−0.34	0.65	0.63	−0.61	0.67
	p	0.001	0.02	0.092	0.005	0.03	0.04	0.011
MDA	*r*		−0.811	−0.44	0.88	0.67	−0.74	0.75
	p		0.01	0.067	0.005	0.01	0.01	0.01
GSH	*r*			0.32	−0.82	0.85	0.67	−0.71
	p			0.21	0.005	0.001	0.004	0.003
CAT	*r*				−0.13	−0.32	0.64	−0.67
	p				0.62	0.21	0.01	0.009
Caspase-3	*r*					0.68	−0.62	0.68
	p					0.01	0.04	0.03
β-catenin	*r*						0.63	0.62
	p						0.01	0.01
Hsp70	*r*							0.62
	p							0.011

MDA = malondialdehyde, GSH = reduced glutathione concentration, CAT = catalase activity, Hsp70 = heat shock protein 70. *r*, Pearson's correlation coefficient. $p < 0.05$ is considered significant.

4. Discussion

The main findings in the present study are: (a) treatment with pentylenetetrazole (PTZ) caused full kindling of rats, enhanced redox state and upregulated the expression of LC3, Hsp70, caspase-3 and β-catenin in the rat hippocampal CA3 region; (b) chronic daily treatment with L-Car (100 mg/kg) caused significant attenuation in seizure stage and duration and reversed the PTZ-induced alterations in the redox sate and the expression of caspase-3, β-catenin and more upregulation of Hsp 70 expression in the rat hippocampal CA3 region.

Acetyl L-Car is a well-known potent antioxidant and has several properties that may suggest its antiepileptic effect, such as protection against glutamate toxicity [19], a favorable effect on energy homeostasis [20–22] and lowering oxidative damage and improving mitochondrial function [23]. Moreover, pretreatment with L-Car before administration of a single convulsive dose of PTZ to mice prolonged latency to seizures and reduced the frequency of tonic-clonic seizures in a dose-dependent manner [24]. On the other hand, Smeland et al. [14] reported in a model of chronic epilepsy (PTZ-kindled epileptic rat) that L-Car failed to slow down the progression and latency to clonic convulsions. However, they reported that it normalized some metabolic parameters in brain tissues such as lactate, (3–13C) alanine, dopamine, *myo*-inositol and succinate in the cortex.

In the present study, the kindling model (a chronic widely-accepted animal model of epilepsy produced by chemical or electrical stimuli) is implicated; of the chemical stimuli, pentylenetetrazole (PTZ) (a gamma amino butyric acid (GABA) A receptor antagonist) produces severe convulsions when administered to animals by using subconvulsive doses that are applied intermittently and repetitively to produce full-blown convulsions [25]. The present study reported a significant increase in seizure score and duration with significant reduction in latency in the PTZ group. These findings are in line with our previous work from our group [18] and others [26,27]. Furthermore, the present study demonstrated that treatment with L-Car in the PTZ-kindling model greatly attenuated tonic-clonic convulsions and the duration of these convulsions, as well as delayed the seizure onset.

It has been demonstrated that reactive oxygen species play a role in the development and progression of epileptic seizure [18,28]. During convulsive seizure, the low antioxidant defense systems can predispose the brain to oxidative stress. Furthermore, the hippocampus may be particularly sensitive to oxidative stress because it has low endogenous levels of vitamin E, an important biochemical antioxidant, relatively to other brain regions [29]. The present study demonstrated a high oxidative stress state in the CA3 region of the hippocampus of PTZ kindled rats in the form a significant increase in MDA (marker of lipid peroxidation) and a significant reduction in antioxidants such as catalase and reduced glutathione (GSH). These findings are in line with previous studies [18,27,30] and suggest that the induction of seizure produces a state of oxidative stress. Furthermore, our results provide good evidence for the role of ROS in the pathophysiology of seizure in this model.

In the present study, co-treatment with L-Car significantly reduced MDA level and increased GSH and catalase levels relative to the PTZ group. In agreement with these findings, previous studies reported that L-car attenuated the seizure duration and severity in pilocarpine-induced epilepsy though its antioxidant effects, which involved the suppression of MDA and upregulation of CAT and SOD activities [31]. These findings lend more evidence to the hypothesis that the antiepileptic effect of L-Car could be partially related to its antioxidant effects.

Autophagy is a regulated process in which the intracellular cellular proteins aggregate and damaged organelles are degraded by the cell lysosomes [32]. LC3 is considered the most reliable cellular marker for autophagy activation [33]. Therefore, we evaluated in the present study LC3 protein expression in the CA3 region of hippocampus by immunohistochemistry. According to our results, little expression of LC3 in the CA3 region in the saline negative control group was demonstrated, while the PTZ group showed marked LC3 expression in brains. Furthermore, there was a positive correlation between LC3 expression and seizure stage, suggesting a role for autophagy in PTZ-induced epilepsy. In agreement with these findings, Shacka et al. [34] showed significant accumulation of LC3-positive autophagy vacuoles in the hippocampus of kainite-induced epileptic mice. Furthermore, Cao et al. [35] demonstrated the presence of autophagy in pilocarpine-induced status epilepticus (SE) models. In addition, Scherz-Shouval and Elazar [36] and Scherz-Shouval et al. [37] hypothesized that reactive oxygen species are the most important activators of autophagy. In the present study, we reported a positive correlation between autophagy and MDA level, which is in line with the previous hypothesis. Moreover, in the present study, we demonstrated for the first time that L-Car inhibits autophagy in PTZ-induced chronic epilepsy.

The role of apoptosis in epilepsy was demonstrated in many previous studies and in different animal models. Sudha et al. [28] and Simonian et al. [38] demonstrated activation of apoptosis in hippocampal regions of KA-induced epilepsy. Moreover, Naseer et al. [39] showed a significant increase in caspase-3 expression with activation of neuronal apoptosis in PTZ-induced epilepsy in adult rats. In line with these previous studies, we demonstrated a significant increase in caspase-3 expression in the CA3 region of hippocampus of the PTZ group when compared to the saline control group. Moreover, treatment with L-Car decreased apoptosis significantly in the hippocampus CA3 region compared to the PTZ group. Activation of apoptosis in PTZ-induced epilepsy could be due to oxidative stress and generation of reactive oxygen species, which disrupt mitochondrial membrane potential and activate the mitochondrial pathway for apoptosis [40]. These findings, along with others suggest a potential role for apoptosis in epilepsy and possible anti-apoptotic effects for L-Car in the epileptic rat model.

Heat shock proteins (HSPs) are stress-induced proteins that play an important role in cellular responses to stress [41]. The expression of HSPs has been detected in different types of cells in the nervous system including neuralgia, neurons and endothelial cells [42]. Previous studies examined the expression of Hsp70 in epilepsy and reported direct relationship between seizure frequency, duration, intensity and Hsp70 expression in both animal models [18,43,44] and human epilepsy [45,46]. It was suggested that the expression of Hsp70 has a protective role for brain cells during epilepsy. The present study demonstrated significant upregulation in Hsp70 expression in the hippocampal CA3 region with positive correlations with the seizure stage. Moreover, we demonstrated that L-Car treatment caused upregulation in Hsp70 in hippocampus, which was reflected in improvement of behavioral changes. These results suggest that upregulation of Hsp70 might be one of the potential mechanism of the antiepileptic effects of L-Car. It is plausible to assume that the highly expressed Hsp70 could protect neurons against oxidative stress and apoptosis. In accordance with this assumption, Li et al. [47] and Zhao et al. [48] reported that the highly inducible Hsp70 has a neuroprotective role in hindering apoptosis, possibly through interacting with p53, which elicits the apoptotic process [49,50]. Furthermore, Kanitkara and Bhonde [51] reported that Hsp-70 can reduce oxidative stress in beta cells and increase glucose-induced insulin release. Moreover, Ayala and Tapia demonstrated that induction of Hsp70 protects the hippocampal neurodegeneration via modulating endogenous glutamate expression [52]. A previous study has shown increased hippocampal CA1 glutamate release in PTZ-treated animals that was inhibited by Chai-Long-Ku-Li-Tan, a Chinese herbal medicine, as a mechanism of its anticonvulsant properties [53]. Examining the effect of L-Car on glutamatergic neurotransmission is highly indicated, however beyond the scope of the current study, and requires further investigation.

In the development of the nervous system, the Wnt-β-catenin signaling pathway plays an important role in neurogenesis, neural differentiation, synapse development and plasticity [54,55]. The Wnt signaling pathway might be involved in a number of CNS disorders such as Alzheimer's disease, schizophrenia and mood disorders [56–58]. Regarding epilepsy, the role of the Wnt-β-catenin signaling pathway showed controversies. Some studies showed upregulation in β-catenin in epileptic animal models such as pilocarpine-induced status epilepticus [59], electroconvulsive seizures rat [60] and Theihaber et al. [61], who reported upregulation of β-catenin in the hypoxic rat model. In addition, Xing et al. [62] demonstrated that β-catenin may not be involved in the development of hippocampal sclerosis of mesial temporal lobe epilepsy. On the other hand, other studies demonstrated downregulation of β-catenin during epileptic animal models such as the Kainite-induced rat model [63,64]. Moreover, Campos et al. [65] demonstrated that β-catenin knockout mice have high seizure susceptibility to PTZ. In the present study, we demonstrated upregulation of β-catenin in CA3 hippocampal regions in the PTZ group, which could be in response to oxidative stress. Downregulation of β-catenin in the L-Car group compared to the PTZ group supports this hypothesis.

5. Conclusions

The results of the present study confirmed the neuroprotective and antiepileptic action of L-Car against PTZ-induced epilepsy. These effects might be explained on the basis of detected L-Car actions such as antioxidant activity, anti-apoptotic effects, suppression of autophagy and upregulation of neuroprotective heat shock protein. However, further knockout experiments will be required to study the detailed molecular mechanism of L-Car.

Acknowledgments: The authors thank Azza Abdelaziz for helping us with the histopathological examination.

Author Contributions: Abdelaziz M. Hussein: shared in the idea, the induction of epilepsy model, biochemical analysis, data analysis and paper writing. Mohamed Adel: shared in the idea, the induction of epilepsy model and paper writing. Mohamed El-Mesery: Western blotting, biochemical analysis and paper writing Khaled M. Abbas: the induction of epilepsy model and sample collections. Amr N. Ali: the induction of epilepsy model and sample collections. Osama A. Abulseoud: shared in the idea, data analysis and paper writing and editing.

Conflicts of Interest: The authors declare no conflict of interest.

References

1. Forsgren, L.; Beghi, E.; Oun, A.; Sillanpaa, M. The epidemiology of epilepsy in Europe—A systematic review. *Eur. J. Neurol.* **2005**, *12*, 245–253. [CrossRef] [PubMed]

2. Hauser, W.A.; Annegers, J.F.; Rocca, W.A. Descriptive epidemiology of epilepsy: Contributions of population-based studies from Rochester, Minnesota. *Mayo Clin. Proc.* **1996**, *71*, 576–586. [CrossRef] [PubMed]

3. Eadie, M. Epilepsy-from the Sakikku to hughlings Jackson. *J. Clin. Neurosci.* **1995**, *2*, 156–162. [CrossRef]

4. Fisher, R.S.; Boas, W.V.; Blume, W.; Elger, C.; Genton, P.; Lee, P.; Engel, J. Epileptic seizures and epilepsy: Definitions proposed by the International League Against Epilepsy (ILAE) and the International Bureau for Epilepsy (IBE). *Epilepsia* **2005**, *46*, 470–472. [CrossRef] [PubMed]

5. Engel, J., Jr. Introduction to temporal lobe epilepsy. *Epilepsy Res.* **1996**, *26*, 141–150. [CrossRef]
6. Pitkanen, A.; Lukasiuk, K. Mechanisms of epileptogenesis and potential treatment targets. *Lancet Neurol.* **2011**, *10*, 173–186. [CrossRef]
7. Cunliffe, V.T.; Baines, R.A.; Giachello, C.N.; Lin, W.H.; Morgan, A.; Reuber, M.; Russell, C.; Walker, M.C.; Williams, R.S. Epilepsy research methods update: Understanding the causes of epileptic seizures and identifying new treatments using non-mammalian model organisms. *Seizure* **2015**, *24*, 44–51. [CrossRef] [PubMed]
8. Tang, F.; Hartz, A.M.S.; Bauer, B. Drug-Resistant Epilepsy: Multiple Hypotheses, Few Answers. *Front. Neurol.* **2017**, *8*, 301. [CrossRef] [PubMed]
9. Levira, F.; Thurman, D.J.; Sander, J.W.; Hauser, W.A.; Hesdorffer, D.C.; Masanja, H.; Odermatt, P.; Logroscino, G. Premature mortality of epilepsy in low- and middle-income countries: A systematic review from the Mortality Task Force of the International League Against Epilepsy. *Epilepsia* **2017**, *58*, 6–16. [CrossRef] [PubMed]
10. Zaccara, G.; Giannasi, G.; Oggioni, R.; Rosati, E.; Tramacere, L.; Palumbo, P.; Convulsive Status Epilepticus Study Group of the Uslcentro Toscana, Italy. Challenges in the treatment of convulsive status epilepticus. *Seizure* **2017**, *47*, 17–24. [CrossRef] [PubMed]
11. Kwan, P.; Brodie, M.J. Early identification of refractory epilepsy. *N. Engl. J. Med.* **2000**, *342*, 314–319. [CrossRef] [PubMed]
12. Bartlett, K.; Eaton, S. Mitochondrial beta-oxidation. *Eur. J. Biochem.* **2004**, *271*, 462–469. [CrossRef] [PubMed]
13. Kido, Y.; Tamai, I.; Ohnari, A.; Sai, Y.; Kagami, T.; Nezu, J.; Nikaido, H.; Hashimoto, N.; Asano, M.; Tsuji, A. Functional relevance of carnitine transporter OCTN2 to brain distribution of L carnitine and acetyl-L-carnitine across the blood-brain barrier. *J. Neurochem.* **2001**, *79*, 959–969. [CrossRef] [PubMed]
14. Smeland, O.S.; Meisingset, T.W.; Sonnewald, U. Dietary supplementation with acetyl-L-carnitine improves brain energy metabolism in healthy mice and increases noradrenaline and serotonin content. *Neurochem. Int.* **2012**, *61*, 100–107. [CrossRef] [PubMed]
15. Hansen, S.L.; Nielsen, A.H.; Knudsen, K.E.; Artmann, A.; Petersen, G.; Kristiansen, U.; Hansen, H.S. Ketogenic diet is antiepileptogenic in pentylenetetrazole kindled mice and decrease levels of N-acylethanolamines in hippocampus. *Neurochem. Int.* **2009**, *54*, 199–204. [CrossRef] [PubMed]
16. Couturier, A.; Ringseis, R.; Mooren, F.C.; Kruger, K.; Most, E.; Eder, K. Carnitine supplementation to obese Zucker rats prevents obesity-induced type II to type I muscle fiber transition and favors an oxidative phenotype of skeletal muscle. *Nutr. Metab. (Lond.)* **2013**, *10*, 48. [CrossRef] [PubMed]
17. Moloney, T.C.; Dockery, P.; Windebank, A.J.; Barry, F.P.; Howard, L.; Dowd, E. Survival and immunogenicity of mesenchymal stem cells from the green fluorescent protein transgenic rat in the adult rat brain. *Neurorehabil. Neural Repair* **2010**, *24*, 645–656. [CrossRef] [PubMed]
18. Hussein, A.M.; Abbas, K.M.; Abuoleseod, O. Effects of Ferulic Acid on Oxidative Stress, Heat Shock Protein 70, Connexin 43 and Monoamines in Hippocampus of Pentylenetetrazole-Kindled Rats. *J. Physiol. Pharmacol.* **2017**, *97*, 579–585. [CrossRef] [PubMed]
19. Forloni, G.; Angeretti, N.; Smiroldo, S. Neuroprotective activity of acetyl-L-carnitine: Studies in vitro. *J. Neurosci. Res.* **1994**, *37*, 92–96. [CrossRef] [PubMed]
20. Aureli, T.; Miccheli, A.; Ricciolini, R.; Di Cocco, M.E.; Ramacci, M.T.; Angelucci, L.; Ghirardi, O.; Conti, F. Aging brain: Effect of acetyl-L-carnitine treatment on rat brain energy and phospholipid metabolism. A study by ^{31}P and ^{1}H NMR spectroscopy. *Brain Res.* **1990**, *526*, 108–112. [CrossRef]
21. Aureli, T.; Di Cocco, M.E.; Puccetti, C.; Ricciolini, R.; Scalibastri, M.; Miccheli, A.; Manetti, C.; Conti, F. Acetyl-L-carnitine modulates glucose metabolism and stimulates glycogen synthesis in rat brain. *Brain Res.* **1998**, *796*, 75–81. [CrossRef]
22. Scafidi, S.; Fiskum, G.; Lindauer, S.L.; Bamford, P.; Shi, D.; Hopkins, I.; McKenna, M.C. Metabolism of acetyl-L-carnitine for energy and neurotransmitter synthesis in the immature rat brain. *J. Neurochem.* **2010**, *114*, 820–831. [CrossRef] [PubMed]
23. Liu, J.; Killilea, D.W.; Ames, B.N. Age-associated mitochondrial oxidative decay: Improvement of carnitine acetyltransferase substrate-binding affinity and activity in brain by feeding old rats acetyl-L-carnitine and/or R-alpha-lipoic acid. *Proc. Natl. Acad. Sci. USA* **2002**, *99*, 1876–1881. [CrossRef] [PubMed]
24. Yu, Z.; Iryo, Y.; Matsuoka, M.; Igisu, H.; Ikeda, M. Suppression of pentylenetetrazol-induced seizures by carnitine in mice. *Naunyn-Schmiedeberg's Arch. Pharmacol.* **1997**, *355*, 545–549. [CrossRef]

25. Dhir, A. Pentylenetetrazol (PTZ) Kindling Model of Epilepsy. *Curr. Protoc. Neurosci.* **2012**, *58*, 9–37. [CrossRef]

26. Rajabzadeh, A.; Bideskan, A.E.; Fazelm, A.; Sankian, M.; Rafatpanah, H.; Haghir, H. The effect of PTZ-induced epileptic seizures on hippocampal expression of PSA-NCAM in offspring born to kindled rats. *J. Biomed. Sci.* **2012**, *19*, 56. [CrossRef] [PubMed]

27. Saha, L.; Bhandari, S.; Bhatia, A.; Banerjee, D.; Chakrabarti, A. Anti-kindling Effect of Bezafibrate, a Peroxisome Proliferator-activated Receptors Alpha Agonist, in Pentylenetetrazole Induced Kindling Seizure Model. *J. Epilepsy Res.* **2014**, *4*, 45–54. [CrossRef] [PubMed]

28. Sudha, K.; Rao, A.V.; Rao, A. Oxidative stress and antioxidants in epilepsy. *Clin. Chim. Acta.* **2001**, *303*, 19–24. [CrossRef]

29. Danjo, S.; Ishihara, Y.; Watanabe, M.; Nakamura, Y.; Itoh, K. Pentylenetrazole-induced loss of blood-brain barrier integrity involves excess nitric oxide generation by neuronal nitric oxide synthase. *Brain Res.* **2013**, *1530*, 44–53. [CrossRef] [PubMed]

30. Shehata, A.M. Neurophysiological Studies on the Effect of Acetone on Pentylenetetrazole-Induced Seizure in Rats. *Bull. Egypt. Soc. Physiol. Sci.* **2011**, *31*, 135–146.

31. Ahmed, M.F.; Mahmoud, M.A. Effect of L-Carnitine on Pilocarpine-Induced Seizures in Rats. *J. Am. Sci.* **2012**, *8*, 612–618.

32. Zhao, X.; Liu, G.; Shen, H.; Gao, B.; Li, X.; Fu, J.; Zhou, J.; Ji, Q. Liraglutide inhibits autophagy and apoptosis induced by high glucose through GLP-1R in renal tubular epithelial cells. *Int. J. Mol. Med.* **2015**, *35*, 684–692. [CrossRef] [PubMed]

33. Wang, Y.; Han, R.; Liang, Z.Q.; Wu, J.C.; Zhang, X.D.; Gu, Z.L.; Qin, Z.H. An autophagic mechanism is involved in apoptotic death of rat striatal neurons induced by the non-N-methyl-Daspartate receptor agonist kainic acid. *Autophagy* **2008**, *4*, 214–226. [CrossRef] [PubMed]

34. Shacka, J.J.; Lu, J.; Xie, Z.L.; Uchiyama, Y.; Roth, K.A.; Zhang, J. Kainic acid induces early and transient autophagic stress in mouse hippocampus. *Neurosci. Lett.* **2007**, *414*, 57–60. [CrossRef] [PubMed]

35. Cao, L.; Xu, J.; Lin, Y.; Zhao, X.; Liu, X.; Chi, Z. Autophagy is upregulated in rats with status epilepticus and partly inhibited by vitamin E. *Biochem. Biophys. Res. Commun.* **2009**, *379*, 949–953. [CrossRef] [PubMed]

36. Scherz-Shouval, R.; Elazar, Z. ROS, mitochondria and the regulation of autophagy. *Trends Cell Biol.* **2007**, *17*, 422–427. [CrossRef] [PubMed]

37. Scherz-Shouval, R.; Shvets, E.; Fass, E.; Shorer, H.; Gil, L.; Elazar, Z. Reactive oxygen species are essential for autophagy and specifically regulate the activity of Atg4. *EMBO J.* **2007**, *26*, 1749–1760. [CrossRef] [PubMed]

38. Simonian, N.A.; Getz, R.L.; Leveque, J.C.; Konradi, C.; Coyle, J.T. Kainic acid induces apoptosis in neurons. *Neuroscience* **1996**, *75*, 1047–1055. [CrossRef]

39. Naseer, M.I.; Ullah, I.; Ullah, N.; Lee, H.Y.; Cheon, E.W.; Chung, J.; Kim, M.O. Neuroprotective effect of vitamin C against PTZ induced apoptotic neurodegeneration in adult rat brain. *Pak. J. Pharm. Sci.* **2011**, *24*, 263–268. [PubMed]

40. Green, D.R.; Reed, J.C. Mitochondria and apoptosis. *Science* **1998**, *281*, 1309–1312. [CrossRef] [PubMed]

41. Soti, C.; Nagy, E.; Giricz, Z.; Vígh, L.; Csermely, P.; Ferdinandy, P. Heat shock proteins as emerging therapeutic targets. *Br. J. Pharmacol.* **2005**, *146*, 769–780. [CrossRef] [PubMed]

42. Foster, J.A.; Brown, I.R. Differential induction of heat shock mRNA in oligodendrocytes, microglia, and astrocytes following hyperthermia. *Brain Res. Mol. Brain Res.* **1997**, *45*, 207–218. [CrossRef]

43. Gass, P.; Prior, P.; Kiessling, M. Correlation between seizure intensity and stress protein expression after limbic epilepsy in the rat brain. *Neuroscience* **1995**, *65*, 27–36. [CrossRef]

44. Zhang, X.; Boulton, A.A.; Yu, P.H. Expression of heat shock protein-70 and limbic seizure-induced neuronal death in the rat brain. *Eur. J. Neurosci.* **1996**, *8*, 1432–1440. [CrossRef] [PubMed]

45. Thom, M.; Seetah, S.; Sisodiya, S.; Koepp, M.; Scaravilli, F. Sudden and unexpected death in epilepsy (SUDEP): Evidence of acute neuronal injury using HSP-70 and c-Jun immunohistochemistry. *Neuropathol. Appl. Neurobiol.* **2003**, *29*, 132–143. [CrossRef]

46. Dericioglu, N.; Soylemezoglu, F.; Gursoy-Ozdemir, Y.; Akalan, N.; Saygi, S.; Dalkara, T. Cell death and survival mechanisms are concomitantly active in the hippocampus of patients with mesial temporal sclerosis. *Neuroscience* **2013**, *237*, 56–65. [CrossRef] [PubMed]

47. Li, W.X.; Chen, C.H.; Ling, C.C.; Li, G.C. Apoptosis in heat induced cell killing: The protective role of hsp-70 and the sensitization eVect of the c-myc gene. *Radiat. Res.* **1996**, *145*, 324–330. [CrossRef] [PubMed]

48. Zhao, Z.G.; Ma, Q.Z.; Xu, C.X. Abrogation of heat-shock protein (HSP) 70 expression induced cell growth inhibition and apoptosis in human androgen-independent prostate cancer cell line PC-3m. *Asian J. Androl.* **2004**, *6*, 319–324. [PubMed]

49. Lee, C.S.; Montebello, J.; Rush, M.; Georgiou, T.; Wawryk, S.; Rode, J. Overexpression of heat shock protein (HSP) 70 associated with abnormal p53 expression in cancer of the pancreas. *Zentralbl. Pathol.* **1994**, *140*, 259–264. [PubMed]

50. Merrick, B.A.; He, C.; Witcher, L.L.; Patterson, R.M.; Reid, J.J.; Pence-Pawlowski, P.M.; Selkirk, J.K. HSP binding and mitochondrial localization of p53 protein in human HT1080 and mouse C3H10T1/2 cell lines. *Biochim. Biophys. Acta* **1996**, *1297*, 57–68. [CrossRef]

51. Kanitkara, M.; Bhonde, R.R. Curcumin treatment enhances islet recovery by induction of heat shock response proteins, Hsp70 and heme oxygenase-1, during cryopreservation. *Life Sci.* **2008**, *82*, 182–189. [CrossRef] [PubMed]

52. Ayala, G.X.; Tapia, R. HSP70 expression protects against hippocampal neurodegeneration induced by endogenous glutamate in vivo. *Neuropharmacology* **2008**, *55*, 1383–1390. [CrossRef] [PubMed]

53. Wu, H.M.; Huang, C.C.; Li, L.H.; Tsai, J.J.; Hsu, K.S. The Chinese herbal medicine Chai-Hu-Long-Ku-Mu-Li-Tan (TW-001) exerts anticonvulsant effects against different experimental models of seizure in rats. *Jpn. J. Pharmacol.* **2000**, *82*, 247–260. [CrossRef] [PubMed]

54. Lie, D.C.; Colamarino, S.A.; Song, H.J.; Desire, L.; Mira, H.; Consiglio, A.; Lein, E.S.; Jessberger, S.; Lansford, H.; Dearie, A.R.; et al. Wnt signalling regulates adult hippocampal neurogenesis. *Nature* **2005**, *437*, 1370–1375. [CrossRef] [PubMed]

55. Wisniewska, M.B. Physiological role of beta-catenin/TCF signaling in neurons of the adult brain. *Neurochem. Res.* **2013**, *38*, 1144–1155. [CrossRef] [PubMed]

56. De Ferrari, G.V.; Papassotiropoulos, A.; Biechele, T.; Wavrant De-Vrieze, F.; Avila, M.E.; Major, M.B.; Myers, A.; Sáez, K.; Henríquez, J.P.; Zhao, A.; et al. Common genetic variation within the low-density lipoprotein receptor-related protein 6 and late-onset Alzheimer's disease. *Proc. Natl. Acad. Sci. USA* **2007**, *104*, 9434–9439. [CrossRef] [PubMed]

57. Lovestone, S.; Guntert, A.; Hye, A.; Lynham, S.; Thambisetty, M.; Ward, M. Proteomics of Alzheimer's disease: Understanding mechanisms and seeking biomarkers. *Expert Rev. Proteom.* **2007**, *4*, 227–238. [CrossRef] [PubMed]

58. Zhang, Y.; Yuan, X.; Wang, Z.; Li, R. The canonical Wnt signaling pathway in autism. *CNS Neurol. Disord. Drug Targets* **2014**, *13*, 765–770. [CrossRef] [PubMed]

59. Fasen, K.; Beck, H.; Elger, C.E.; Lie, A.A. Differential regulation of cadherins and catenins during axonal reorganization in the adult rat CNS. *J. Neuropathol. Exp. Neurol.* **2002**, *61*, 903–913. [CrossRef] [PubMed]

60. Madsen, T.M.; Newton, S.S.; Eaton, M.E.; Russell, D.S.; Duman, R.S. Chronic electroconvulsive seizure up-regulates betacatenin expression in rat hippocampus: Role in adult neurogenesis. *Biol. Psychiatry* **2003**, *54*, 1006–1014. [CrossRef]

61. Theilhaber, J.; Rakhade, S.N.; Sudhalter, J.; Kothari, N.; Klein, P.; Pollard, J.; Jensen, F.E. Gene expression profiling of a hypoxic seizure model of epilepsy suggests a role for mTOR and Wnt signaling in epileptogenesis. *PLoS ONE* **2013**, *8*, e74428. [CrossRef] [PubMed]

62. Xing, X.L.; Sha, L.Z.; Zhang, D.; Shen, Y.; Wu, L.W.; Xu, Q. Role of beta-catenin in the pathogenesis of mesial temporal lobe epilepsy. *Zhongguo Yi Xue Ke Xue Yuan Xue Bao* **2011**, *33*, 659–662. [PubMed]

63. Busceti, C.L.; Biagioni, F.; Aronica, E.; Riozzi, B.; Storto, M.; Battaglia, G.; Giorgi, F.S.; Gradini, R.; Fornai, F.; Caricasole, A.; et al. Induction of the Wnt inhibitor, Dickkopf-1, is associated with neurodegeneration related to temporal lobe epilepsy. *Epilepsia* **2007**, *48*, 694–705. [CrossRef] [PubMed]

64. Goodenough, S.; Schleusner, D.; Pietrzik, C.; Skutella, T.; Behl, C. Glycogen synthase kinase 3beta links neuroprotection by 17 beta-estradiol to key Alzheimer processes. *Neuroscience* **2005**, *132*, 581–589. [CrossRef] [PubMed]

65. Campos, V.E.; Du, M.; Li, Y. Increased seizure susceptibility and cortical malformation in beta-catenin mutant mice. *Biochem. Biophys. Res. Commun.* **2004**, *320*, 606–614. [CrossRef] [PubMed]

brain
sciences

MDPI

Review

Diagnosis and Surgical Treatment of Drug-Resistant Epilepsy

Chinekwu Anyanwu [1] and Gholam K. Motamedi [2],*

[1] Department of Neurology, Virginia Tech Carilion School of Medicine, Roanoke, VA 24016, USA; nekwusky@gmail.com

[2] Department of Neurology, Georgetown University Medical Center, Washington, DC 20007, USA

* Correspondence: motamedi@georgetown.edu; Tel.: +1-202-444-4564; Fax: +1-877-245-1499

Received: 24 November 2017; Accepted: 16 March 2018; Published: 21 March 2018

Abstract: Despite appropriate trials of at least two antiepileptic drugs, about a third of patients with epilepsy remain drug resistant (intractable; refractory). Epilepsy surgery offers a potential cure or significant improvement to those with focal onset drug-resistant seizures. Unfortunately, epilepsy surgery is still underutilized which might be in part because of the complexity of presurgical evaluation. This process includes classifying the seizure type, lateralizing and localizing the seizure onset focus (epileptogenic zone), confirming the safety of the prospective brain surgery in terms of potential neurocognitive deficits (language and memory functions), before devising a surgical plan. Each one of the above steps requires special tests. In this paper, we have reviewed the process of presurgical evaluation in patients with drug-resistant focal onset epilepsy.

Keywords: medically intractable epilepsy; EEG; epilepsy surgery; MRI; epilepsy; seizures

1. Introduction

The majority of patients with epilepsy can become seizure-free with antiepileptic drugs (AEDs) but about a third will continue to have seizures. However, patients with drug-resistant focal onset epilepsy (the most common type of epilepsy), may become seizure-free, or have significant seizure reduction, through epilepsy surgery. In particular, the efficacy and safety of anterior temporal lobectomy in patients with temporal lobe epilepsy has been established through randomized controlled trials [1].

Despite the evidence of its superiority to medical therapy in patients with drug-resistant epilepsy (DRE), epilepsy surgery is still underutilized as only a small group of these patients undergo presurgical evaluation and even a smaller group end up having surgery. It is still a common practice to try almost all available AEDs before referring patients for possible epilepsy surgery, while with proper management, diagnosis of DRE can be established within 1–2 years after seizure onset.

Currently, out of about one million patients with DRE in the United States—the majority of whom are suffering from focal onset epilepsy and potential candidates for surgery—less than 1% are referred to epilepsy centers and only 2000 to 3000 undergo surgery. The recommended approach is to refer patients who have failed adequate trials of two AEDs, either as monotherapy or in combination, to an epilepsy center for presurgical evaluation [2–4].

This article will review proper management of patients with DRE, presurgical evaluation, and different surgical treatment options based on the current recommendations by the international league against epilepsy (IALE), and established guidelines.

2. New Classification of Seizure Types

In 2017, the international league against epilepsy (ILAE) released a new classification of seizure types. This revision was based on the previous classification that has been in use since 1981. The new classification was developed to reflect the developments in our understanding of the disease since then, in particular in brain imaging, electrophysiology, and genetics. The main differences in the new classification include listing of certain new focal seizure types that used to be under the category of generalized seizures, replacing the term consciousness with awareness in description of seizures, and classifying focal seizures based on the first clinical manifestation (except for altered awareness), adding new types of generalized seizures, and adding or changing certain terms to clarity the terminology. In particular, the most common type of epilepsy, which is also potentially treatable through surgery, complex partial epilepsy, will be referred to as focal onset epilepsy with loss of awareness [5,6].

3. Drug-Resistant (Refractory) Epilepsy

It has been shown that 47% of patients with new-onset epilepsy can achieve complete seizure-freedom with the first AED, 13% of the remaining 53% may become seizure-free using a second agent, and only 4% with a third agent and/or polytherapy [7]. Therefore, failure of two AEDs, either as monotherapy or combination therapy, would meet the criteria for DRE [8]. However, it should be emphasized that the AEDs must have been tried adequately. An "adequate trial" includes, choosing an appropriate AED for the type of seizure being treated (e.g., failure of carbamazepine in treating idiopathic generalized epilepsy would not be considered a true failure), and titrating up the dose to the maximum tolerated dose. An AED that triggers allergic reaction or causes significant adverse effects that requires switching to another AED, may not be counted as failure either.

It is also critical to define "seizure-freedom" when assessing the success or failure of an AED. The recommended definition is sustained seizure freedom for a period 3 times the longest inter-seizure interval, or 1 year, whichever that is longer [8].

4. Presurgical Evaluation

The process referred to as presurgical evaluation is the most critical aspect of epilepsy surgery. All patients who are suspected of having- or properly diagnosed with- drug-resistant epilepsy should be referred to an epilepsy center for presurgical evaluation. During this process, the patient will be further evaluated to confirm the diagnosis and to classify the seizure type.

This step is critical since close to half of patients with epilepsy have idiopathic generalized seizures which cannot be treated with resective epilepsy surgery. Only focal onset epilepsy is amenable to surgery. Therefore, only those with DRE who have focal seizures with loss of awareness (formerly, complex partial seizures), with or without secondary generalization, are surgical candidates and can complete the process of presurgical evaluation.

It is also crucial to confirm that the patient has true epilepsy and not an epilepsy mimic since patients may seem refractory for presumed seizures but may have non-epileptic spells such as syncope or psychogenic non-epileptic seizures (PNES). Therefore, the initial evaluation includes obtaining proper history, seizure semiology, and electroencephalographic (EEG) findings are role in this process.

The ultimate goal of presurgical evaluation is to identify the epileptogenic zone (region), and to establish the safety of a prospective brain surgery, so that the resection can be performed with minimal functional impairment while achieving seizure freedom, if possible.

Identification of seizure focus involves extensive workup including seizure characterization, lateralization, and localization through detailed evaluation of history, video-EEG (VEEG) monitoring using scalp EEG electrodes or—if indicated—via surgically implanted intracranial electrodes, different neuro-imaging modalities, neuropsychological testing, and if indicated, intracarotid Amobarbital injection procedure (IAP; Wada test).

The fundamental goal in presurgical workup includes defining the epileptogenic zone (epileptogenic region) which is the area sufficient for generation of seizures, and the minimum amount of cortex that must be resected (inactivated or completely disconnected) to produce seizure freedom. A variety of diagnostic tools should be utilized in evaluation of the various cortical zones. They include seizure semiology, VEEG recording, neuro functional testing and neuroimaging techniques with a goal to define the cortical zones involved in seizure generation and propagation. Other cortical zones of interest besides the epileptogenic zone include seizure-onset zone (cortical region that initiates clinical seizures), symptomatogenic zone (where initial ictal semiology is produced), irritative zone (where interictal spikes are generated), and functional deficit zone (cortical area that functions abnormally during interictal period) [9,10].

While the ultimate tool in detecting the epileptogenic zone is the VEEG findings (more accurately via direct intracranial recording), it can be significantly facilitated by high resolution brain magnetic resonance imaging (MRI) (structural integrity), diffusion tensor imaging (cellular integrity), magnetic resonance spectroscopy (metabolite and biochemical data), or through physiological imaging modalities such as positron emission tomography (PET) and single-photon emission computerized tomography (SPECT) scans (glucose or other ligands' metabolism, and cerebral blood flow, respectively). The irritative zone (area that generates interictal epileptiform activities) can be detected by EEG, magnetoencephalography (MEG) and special functional MRI techniques. The seizure-onset zone (area of cortex that initiates clinical seizures) can be measured by scalp/intracranial EEG recording and ictal SPECT.

The symptomatogenic zone (area of cortex which produces the initial ictal symptoms or signs) can be determined by the seizure semiology. The functional deficit zones are measured with neuropsychological examination and functional imaging studies at the baseline (interictal period).

Despite the availability of multiple diagnostic modalities, different cases may require different diagnostic tests therefore, the evaluations process should be tailored to the individual. In particular, since a major classification in epilepsy is differentiating temporal lobe epilepsy (TLE) from extra-temporal epilepsy, it is crucial to make that determination during the presurgical evaluation. This distinction is based on the critical role of temporal lobe in neurocognitive functioning, its unique anatomy, and the clinical evidence regarding its responsiveness to surgery.

In particular, TLE with mesial temporal sclerosis (MTS) may not require extensive diagnostic work up as it can be confirmed with concordant ipsilateral anterior-mid temporal epileptiform discharges without the need for intracranial recording. However, some patients may require intracranial (subdural) monitoring for seizure focus lateralization and/or accurate localization as well as functional cortical mapping. In general, about half of patients with TLE and almost all extra-temporal cases require intracranial EEG recording for both purposes of pinpointing the seizure onset focus and/or cortical mapping to delineate eloquent cortices for safety [11].

It should be noted that while patients who fail two appropriate trials of AEDs should be considered drug-resistant and should undergo presurgical evaluation, they may benefit from adjustments to their current AED regimens as seen appropriate. Therefore, starting the process of presurgical evaluation does not preclude using other AEDs, especially given the available variety of AEDs with different mechanisms of action and side effect profiles.

4.1. Scalp Video-EEG Monitoring

The standard procedure after confirming that the patient has refractory epilepsy usually starts with VEEG monitoring to further characterize, lateralize and possibly localize the seizure focus. Localization of the seizure onset zone through scalp EEG may be limited as it detects epileptiform activity that synchronizes at least 6 cm^2 of the cortex [12]. Deeper seizure foci may not be detected via surface (scalp) electrodes. Further, scalp EEG recording is usually insufficient in extra-temporal epilepsy or even in non-lesional (normal MRI) TLE. Using extra electrodes placed based on the 10-10 international electrode system may add significant accuracy to scalp recording and in some cases avoid intracranial recording (Figure 1).

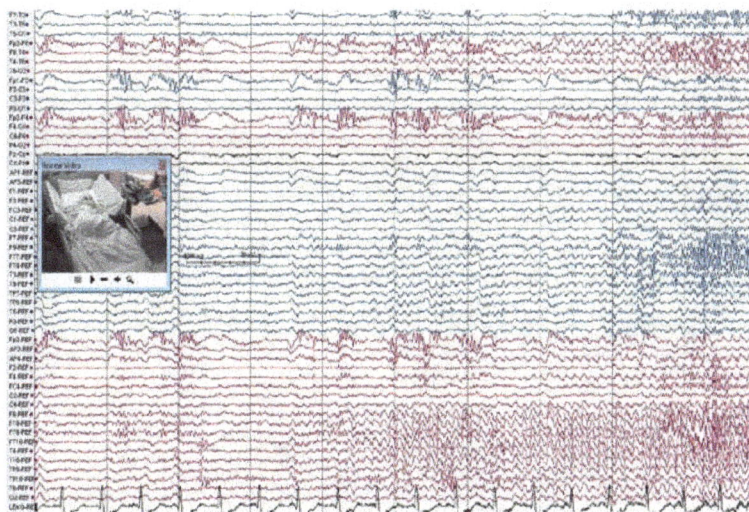

Figure 1. Right temporal lobe seizure onset in a patient with right hippocampal sclerosis and temporal lobe epilepsy (TLE). The EEG shows buildup of rhythmic repetitive sharp theta discharges arising in the right anterior-mid temporal regions. Scalp recording during long-term monitoring using extra electrodes placed according to the 10-10 electrode system for clear distinction between frontal and basal-lateral temporal head regions.

When seizure onset zone is poorly lateralized because of alternating seizure onset lateralization, bitemporal asynchrony, or frequent bilateral epileptiform discharges, it indicates a less favorable postsurgical seizure outcome [13]. Unilateral hippocampal atrophy on MRI and concordant unilateral interictal spikes are highly predictive of concordant ictal localization [14].

Patients are often admitted for several days, depending on their seizure frequency. Most often their AEDs are tapered off in order to capture about 3–5 typical and consistent seizures. There is no consensus on tapering process but in general very rapid tapering is avoided to minimize risk of Status Epilepticus or triggering aberrant seizure onset zones. Upon the completion of scalp VEEG monitoring, some patients may not qualify for surgery; this includes patients with primary generalized seizures and those with clear multifocal seizures, although some in the latter group may qualify for intracranial recording for further elaboration. When scalp EEG cannot adequately localize the seizure focus, particularly in patients with normal brain MRI, intracranial EEG monitoring would be justified.

4.2. Brain Imaging Techniques

Advances in brain imaging technology have substantially improved seizure localization and surgical outcome [12]. Neuroimaging studies are not routinely indicated in all types of epilepsy. For example, patients with idiopathic (primary) generalized epilepsy syndromes such as absence epilepsy and Juvenile myoclonic epilepsy diagnosed clinically and through EEG typically do not require imaging [13]. All patients with clinical and EEG evidence of focal onset epilepsy should have brain imaging which includes brain MRI.

4.2.1. MRI (Magnetic Resonance Imaging)

The principal role of MRI is to define structural abnormalities that may be the cause of seizure disorder. A high resolution brain MRI, usually referred to as Epilepsy Protocol MRI, is recommended as the modality of choice for all patients presenting with symptomatic focal epilepsies, with or without generalization. Detailed MRI sequences may be added to increase diagnostic yield depending on the etiology. Common sequences used by most epilepsy centers include thin-section (1 mm) coronal oblique T1 gradient echo, coronal oblique T2 series, high-resolution 3D sequences (sensitive to subtle cortical dysplasias or small tumors), and T2 FLAIR (fluid attenuated inversion recovery) images, performed on 3 Tesla, or higher, MRI systems [14].

Temporal lobe epilepsy is the most common type of focal epilepsy and the most common MRI finding in these patients is MTS, although this pathology may not be present at the time of seizure onset and may take many years to develop. The main radiological findings in MTS include hippocampal atrophy, internal structural derangement, and T2 hyperintensity as best seen on coronal T2 flair images (Figure 2). Other cortical epileptogenic lesions include focal cortical dysplasia (FCD), tubers, gliosis, infectious process (e.g., neurocysticercosis, abscesses), benign and malignant tumors, and inflammatory processes (Rasmussen's encephalitis). Not all lesions are epileptogenic and patients may present with dual pathologies. In MRI-negative neocortical epilepsies, the most common pathology is FCD type 1 and 2. Type 2 FCDs are preferentially located at the bottom of the sulcus with high-resolution MRI making it possible to visually identify such in an increasing number [15,16].

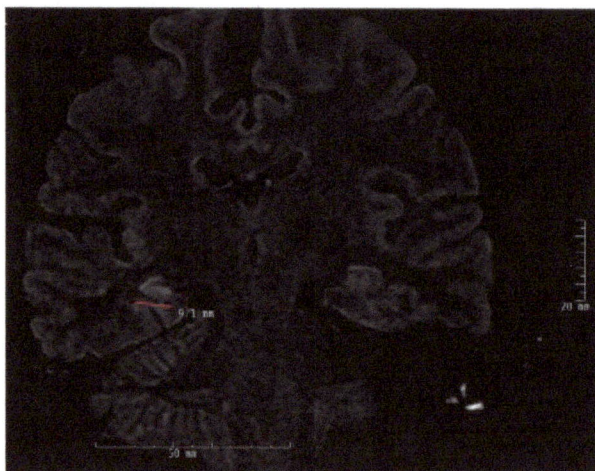

Figure 2. Right mesial temporal sclerosis (MTS). There is atrophy and increased signal intensity in the CA1-3 regions of the right hippocampus. Coronal view; fluid-attenuated inversion recovery (FLAIR) sequence.

4.2.2. SPECT Scan (Single-Photon Emission Computerized Tomography)

Structural lesions do not always correlate with clinical semiology and EEG findings but the regional cerebral blood flow is useful tool to localize these seizures. Ictal SPECT is more sensitive and specific in temporal lobe epilepsies given that the seizures are longer. The strength of SPECT is the ability to obtain images related to regional cerebral blood flow (rCBF) at the time of seizures.

Interictal or ictal SPECT have been used to localize the ictal onset and ictal propagation patterns and add to the evidence of abnormalities in the involved site [15,17]. Meta-analytic sensitivities of SPECT in patients with TLE have been reported as 44% (interictally), 75% (postictally) and 97% (ictally) [18] as opposed to 66% (ictally), and 40% (interictally) in extra-temporal seizures [19,20].

Ictal SPECT studies appear to correlate with outcome when there is close concordance between the area of hyperperfusion and the resected area and vice versa with surgical failure when there is poor concordance. Both ictal and interictal SPECT showing a change from hypoperfusion (interictal period) to hyperperfusion (ictal period) is more reliable than an abnormality in either stage alone. However, it has been shown that the sensitivity of ictal SPECT is higher than that of interictal SPECT because of the large CBF increase from the baseline that occurs during the ictal phase [21]. A significant limitation of SPECT scan is its logistical complexity so that a significant number of major epilepsy centers do not use it.

4.2.3. FDG-PET Scan (Fluorodeoxyglucose Positron Emission Tomography)

Pet scan obtained with FDG (18 F-fluorodeoxyglucose) is considered a non-invasive technique. PET scans demonstrate hypermetabolism or increased glucose uptake during an ictal event and hypometabolism during the interictal period. When MRI is normal, PET scan may be indicated to aid in localization. It is often correlated with other studies including MRI findings, EEG neuropsychological testing and Wada test. Interictal FDG PET is known to be more sensitive than interictal SPECT in localizing extra-temporal epilepsy [22]. The changes, however, are more extensive than the structural and EEG abnormalities and may involve other regions including the ipsilateral suprasylvian and parietal regions. Interictal FDG-PET is considered to be the best imaging technique to assess the functional deficit zone (FDZ).

In TLE, interictal studies show hypometabolic areas in the epileptogenic regions in approximately 70–80% of patients (Figure 3) [23]. Ipsilateral PET hypometabolism showed a predictive value of 86% for good outcome in meta-analysis of 46 studies [24].

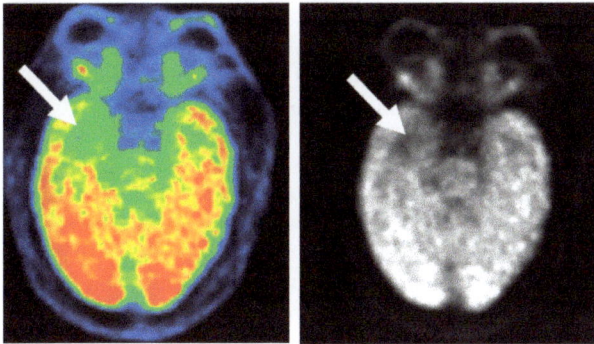

Figure 3. Interictal FDG PET (fluorodeoxyglucose positron emission tomography) scan. Decreased glucose uptake in the right temporal area (arrows) compared to the left side, in a patient with right mesial temporal lobe epilepsy (MTLE).

4.2.4. MEG (Magnetoencephalography)

Magnetoencephalography maps interictal magnetic dipole sources onto MRI to produce a magnetic source image. It is a more recent imaging modality in the presurgical evaluation of focal epilepsy that has proven helpful in the detection of epileptogenic foci even in patients with inconclusive results from other non-invasive test. The additional information provided by these techniques has been demonstrated to help those patients with non-localizing MRI or extra-temporal epilepsy. MEG is in principle more sensitive to neuronal activity from superficial than deep-seated structures. It is more sensitive in detecting currents that are tangential to the surface of the scalp whereas EEG is sensitive to tangential and radial neuronal activities. MEG systems allow rapid high-resolution recordings of cortical function and dysfunction that are neither attenuated nor distorted by the skull or other variable intervening tissue layers between the scalp and brain. It primarily detects the magnetic fields induced by intracellular currents, whereas scalp EEG is sensitive to electrical fields generated by extracellular currents.

The site of surgery in 58 patients were correctly predicted by intracranial EEG in 70%, compared with MEG (52%), interictal scalp VEEG (45%) and ictal scalp VEEG (33%) [25]. MEG has shown equal effectiveness compared to EEG in terms of recording epileptic spikes with a tendency to register epileptic spikes in more patients [26].

4.2.5. fMRI (Functional MRI)

While fMRI has not been established as a valid test to replace Wada memory test [27], it may be a potential complement to standard neuropsychological test to predict postoperative verbal memory outcome. It has been suggested [28,29] that it may be used to determine language dominance prior to temporal lobectomy [30–33]. Overall, besides the evidence indicating its utility as a replacement of Wada language test [34,35], to date, the replacement of the Wada memory test has proved to be more difficult [33,36,37].

The concordance between fMRI and Wada language lateralization has been shown in a number of studies reporting concordance. In one such study in 100 patients with different focal epilepsies, there was 91% concordance between the two tests. The overall rate of false categorization by fMRI was 9%, ranging from 3% in left-sided TLE to 25% in left-sided extra-temporal epilepsy [38].

4.2.6. Intracarotid Amobarbital Injection Procedure (IAP, Wada Test)

A main focus of the presurgical evaluation is to estimate the risk of language- and verbal memory decline in patients undergoing anterior temporal lobectomy (ATL) on the language dominant side (usually, the left). Intracarotid Amobarbital injection procedure, commonly referred to as Wada test (after Juhn Atsushi Wada), is considered the gold standard for pre-operative lateralization of language dominance, and a measure of memory function [39]. The Wada test is also effective in predicting seizure control and degree of verbal memory decline post operatively.

Wada test is indicated in patients who may undergo temporal lobectomy. However, despite the high accuracy of the Wada test in lateralizing language and memory function, this test is associated with false negatives and false positives and needs to be performed by experienced practitioners [40]. In good hands, the test is considered reasonably safe with potential minor and major complications that have been reported to be rare, but in some reports as high as 1.09% [41]. Despite the risk, Wada test clearly remains useful in patients without clear language lateralization or with suspected atypical language lateralization, but more critically in patients at risk of postsurgical memory loss [42].

In some cases the validity of the test has been questioned [43]. Prior to the procedure, cerebral angiography is used to assess the vasculature and extent of cross-over flow to contralateral arteries. There are different anatomical variations that lead to unreliable results. EEG will be continuously recorded throughout the procedure. During the procedure, amobarbital (100–150 mg) is injected into each carotid artery to produce, one at a time, to induce a temporary disruption in function in the ipsilateral temporal lobe while the patient gets involved in a series of different language and memory related tasks. Other agents including, propofol, methohexital and etomidate have also been used [44–46].

Upon injection of the language dominant side, there will be global aphasia, while the patient would have only dysarthria after injecting the non-dominant temporal lobe. The more delicate function of Wada test is to determine the "adequacy" of the contralateral hippocampus, while shedding light on the "reserve" capacity of the ipsilateral side (seizure side). To be able to undergo anterior temporal lobectomy, ideally the patient is required to correctly identify at least 3 more objects out of a total of 8 target objects, after injecting the ipsilateral side ("split" score of 3). This would indicate enough adequacy on the contralateral hippocampus for a reasonable function ipsilateral temporal lobectomy. A suggested partial alternative to Wada test is fMRI for language lateralization however, there are no standardized methods to assess memory using fMRI.

Currently, the test is performed in a subset of patients who appear to be at risk for clinically significant memory loss. Therefore, some patients with TLE, including those with left-sided TLE, may not need Wada test, depending on results of VEEG, MRI, fMRI, neuropsychological and PET scan results. Patients with left-sided TLE and MTS, ipsilateral hypometabolism on PET scan, low baseline performance in verbal memory and naming on neuropsych testing, and left language dominance on fMRI, may not undergo the test. However, those with nonlesional brain MRI, normal PET scan, and normal baseline verbal memory and naming on neuropsych testing may benefit from Wada test [27,47].

4.2.7. Neuropsychological Testing

While Wada test may not be required in all patients undergoing presurgical evaluation cases, the non-invasive outpatient neuropsychological testing should be performed in all cases. It establishes a baseline neurocognitive profile that can be used for postoperative comparison, and also helps with seizure focus lateralization, localization, and postoperative outcome prediction [48–50]. Neuropsychological testing evaluates a variety of neurocognitive features particularly, but not limited to, in patients with TLE. These include verbal memory, visual memory, verbal fluency, and other aspects of cognitive performance. For example, a typical patient with dominant (usually left-sided) epilepsy would have significant language and memory related deficits such as verbal memory and word finding difficulties.

On the other hand, the same type of seizure originating in the non-language dominant side would typically result in visual memory deficits. Therefore, it is crucial to both establish a baseline, and also to corroborate the results of prior work up since neuropsychological test results may further confirm—or argue against—the assumed localization and lateralization of the epileptogenic zone. The testing includes evaluation of pre-morbid intellectual function, memory function, higher executive function, language function and visuo-spartial function.

Memory and language testing contribute to the assessment of the functional integrity of the contralateral lobe. When testing suggests severe memory and language deficiencies in the contralateral hemisphere, the risk of developing postoperative memory and language impairment is high. Memory disturbances have consistently been found in patients with temporal lobe seizures because of the relationship between mesial temporal structures and memory processes [51–55].

5. Intracranial EEG Recording (Intracranial Electroencephalography; iEEG)

In an increasing number of patients EEG data recorded through scalp electrodes fails to clearly lateralize or localize the seizure focus. This mandates, recording the EEG directly from the brain through surgically implanted subdural surface electrodes and/or depth electrodes inserted into the brain tissue. This group of patients include those with non-lesional brain MRI, unclear seizure onset lateralization, uncertain seizure onset localization, possible bilateral or multifocal seizure disorder, suspected dual pathology, discordant noninvasive data (such as interictal and ictal EEG, structural or functional neuroimaging and neuropsychological testing or EEG findings), and seizures likely originating in eloquent cortex requiring accurate localization and cortical mapping prior to a possible resection surgery.

Implanting recording electrodes, either as depth- or grids and strips, comes with its own challenges. The location and number of hardware should be decided on an individual basis. In general, it seems safer to use minimal required hardware, the same principle that applies to AEDs. However, such general principles cannot dictate the exact surgical plan and therefore the decision-making process should be individualized in every case [56,57].

The use of intracranial EEG to guide surgical resections in some patients with TLE is controversial, particularly in those with presumed unilateral mesial temporal lobe foci [58]. Most patients with unilateral hippocampal atrophy with concordant scalp EEG or functional imaging studies have excellent outcomes after standard antero-medial temporal resections without intracranial monitoring with continued postoperative seizures seen in about 30% of patients [59] and some of these may be due to a dual pathology [60]. Patients with bilateral temporal lobe seizure onsets are known to have poorer surgical outcome than those with unilateral foci [61].

5.1. Subdural Electrodes

Subdural electrodes consist of small metal discs embedded in teflon or silastic sheath material and arranged in different configurations. They can be arranged in single column (strips) or in rows and columns (grid) in different sizes and configurations. These strips are implanted subdurally over the surface of brain via burr holes to insert strips, or craniotomy for grids. Large areas of the frontal, temporal, parietal, interhemispheric (para-falcine), or occipital convexities can be covered with appropriately sized grids and strip electrodes within the subdural space directly over the brain (Figure 4; supplementary material).

Given that substantial parts of the cortex are in the depths of the sulci, subdural grids cannot effectively record activities from these areas. Stereotactic EEG (SEEG) on the other hand can assess deep sulci and gyri including mesial brain, orbito-frontal, insular and basal regions implantations without difficulty. However when mapping of the eloquent cortex is required, subdural electrodes are needed, in some occasion along with SEEG [62].

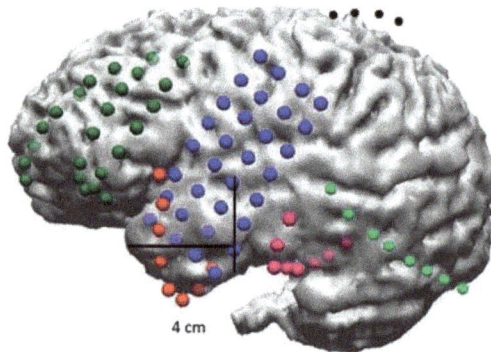

Figure 4. 3D rendering of baseline brain magnetic resonance imaging (MRI) and surgically implanted subdural electrodes (grids and strips).

Besides their ability to record seizures directly from the cortical surface, grid electrodes are the standard method of extraoperative cortical mapping to delineate eloquent cortex. A review of 71 patients who underwent subdural grid placement found transient complications in a majority of patients with two deaths reportedly due to increased intracranial pressure [63]. The authors concluded that there was a relationship between the size of the electrode arrays and the incidence of complications. Recording delayed for more than 2 weeks may increase chances of infection and at times lead malfunction. Epidural electrodes may be used for localization of seizure onset but they are limited by the fact that they can only record from the lateral convexity of the cerebral hemispheres [64].

5.2. Stereoelectroencephalography (SEEG)

Recording EEG signals through surgically implanted depth electrodes provide the best coverage for deeper structures (such as hippocampus, amygdala and insula) and deep sulci. The SEEG method was originally developed by Jean Talairach and Jean Bancaud during the 1950s [65]. The first SEEG implantation was performed on 3 May 1957 at St. Anne in Paris. Its safety, along with the safety of subdural grids, has been well established [66]. The technical aspects remained essentially unchanged for several years but recent advancements has popularized the procedure. Less surgical complications have been observed with SEEG compared to subdural grids.

Risk of infection and intracranial hemorrhage have been reported in 1% and 0.8–1% of patients, respectively. In other series small hemorrhages have been reported in 5.5%, of whom only 0.9% (3/12) required surgery and no mortality were reported [67,68]. In one study, the using SEEG has been shown to confirm the epileptogenic zone in 154 patients (77.0%) with 134 patients undergoing SEEG guided resection of whom 61 patients who were followed up remained seizure free at 12 months [69,70].

The planning of SEEG implantation requires formulating precise hypotheses about the possible epileptogenic zone, seizure onset and propagation zones to be tested. It is also important to understand the functional networks involved in the primary organization of the epileptic activity while formulating a hypothesis. The trajectory passes through several target points including deep sulci, different lobes and gyri orthogonally or obliquely considering different cortical cytoarchitectural areas involved in seizure patterns. For example, in a suspected frontal lobe epilepsy case, a trajectory may pass through primary motor cortex, supplementary motor area, frontal eye field, and deep sulci.

Depth electrodes in various lengths and number of contacts are implanted using conventional stereotactic technique or by the assistance of stereotactic robotic devices through 2.5 mm diameter drill holes.

5.3. Intracranial Seizure Lateralization and Localization

After implanting the intracranial electrodes, the patients are monitored in the epilepsy monitoring unit for 1–2 weeks or longer in some centers. The duration of iEEG recording is determined by seizure frequency, number of seizures needed to make a decision, and the time needed to perform mapping of the eloquent cortex.

Interpretation of the data is based on EEG pattern recognition as well as clinical semiology (symptomatology). An accurate interpretation of intracranial ictal EEG requires appropriate electrode coverage of potential epileptogenic areas. A key point is to have placed the iEEG electrodes in proper locations i.e., at—or in close vicinity to—the epileptogenic zone and areas around it. Besides the more common ictal EEG patterns such as buildup of rhythmic repetitive sharp or spike discharges (Figure 5), there is increasing evidence that high frequency oscillations (HFOs) that may be detected visually at seizure onset, represent the epileptogenic zone. Interpretation may be difficult in some cases since frequency range differs from one cerebral area to another because of neurophysiological properties of the structure and from one etiology to another [71].

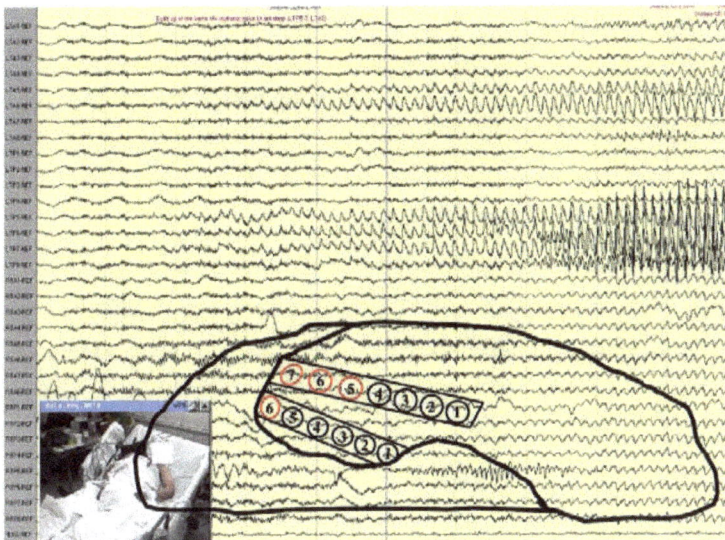

Figure 5. Ictal onset EEG recording through two 1 × 8 subdural strips placed over the basal temporal region in a patient with left temporal lobe epilepsy (TLE). Buildup of rhythmic sharp discharges initiated at a few electrodes (red circles) before rapid propagation to other electrodes and generalization.

Bursts of HFOs may also be recorded during interictal periods and may be a marker of epileptogenicity if the same bursts will develop into sustained fast activity at seizure onset. Sometimes ictal baseline direct current (DC) shifts may precede the fast ictal discharges [72].

A specific interictal pattern in FCD is the occurrence of fast discharges combined with almost continuous interictal spiking in the same region [73]. In localization of seizure onset, the most reliable part is the first clear synchronizing electrical change seen in a limited number of electrodes [74] that precedes the onset of clinical semiology. Mesial temporal lobe seizures may also begin as low-voltage, high-frequency discharge without preictal spiking [75].

5.4. Electrical Stimulation (Cortical) Mapping

The purpose of cortical mapping is to delineate the borders of eloquent cortex. This procedure is performed in patients whose seizure onset zone is located in the vicinity of language or motor cortices. The purpose is to avoid removing such cortical areas hence functional deficits. An alternating polarity square wave pulse pair stimulation (0.3 msec in duration, at 50–100 Hz, lasting 3–5 s) is applied to a pair of electrodes located in areas near the border of presumed eloquent cortex (Figure 6). Caution must be taken when stimulating cortex as a seizure may be provoked particularly when stimulating highly epileptogenic cortical regions such as primary motor cortex and mesial temporal structure. A provoked seizure is usually heralded by after discharges (Figure 7). [76].

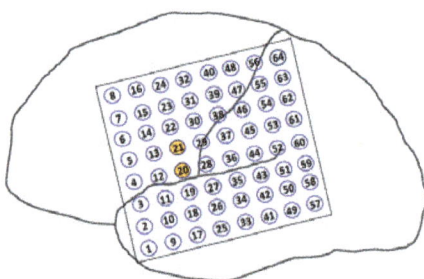

Figure 6. Stimulation cortical mapping. Electrodes 20 and 21, located in the prerolandic motor cortex, are stimulated based on a particular paradigm including slow titration of current density. The patient will be observed for any motor activity in the corresponding body part (face and hand) while engaged in a conversation or reading out loud.

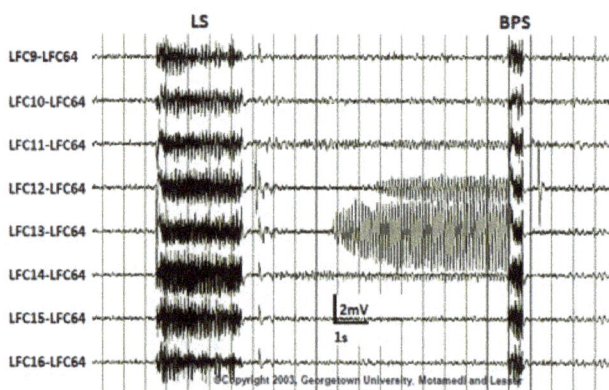

Figure 7. During stimulation mapping the pulse stimulus (LS, localization stimulus) may induce afterdischarges (AD). In some cases, a brief pulse stimulus (BPS) may successfully terminate the AD and prevent a clinical seizure.

6. Epilepsy Surgery

Following the completion of presurgical evaluation and localization of the epileptogenic zone as well as establishing the safety of a resection surgery in the individual patient, the proper surgical option can be decided.

Depending on the seizure type, seizure localization, presence of absence of an identifiable pathology on brain imaging (lesional vs. nonlesional cases, respectively), and patient's functional baseline, different surgical approaches and methods can be used. These include potentially curative surgeries i.e., ATL (using different variations in technique), and resection of the epileptogenic zone (often extra-temporal), or hemispherectomy and palliative surgical approaches such as corpus callosotomy.

Given the higher prevalence of temporal lobe epilepsy and its distinct features i.e., its critical role in basic cognitive functioning and its unique anatomy, epilepsy surgery is commonly categorized and discussed as two separate categories of temporal- vs. extra-temporal.

Temporal lobe epilepsy surgery most often is performed as anterior temporal lobectomy and includes the removal of anterior and mesial structures which naturally raises concerns about postoperative language and memory deficits [1]. Therefore, as discussed earlier, specific diagnostic tests have to be performed to address those concerns before a safe surgical method can be devised.

Extratemporal epilepsy surgery on the other hand, involves the removal of an identified seizure focus as long it does not overlap with eloquent cortex. Generally, well delineated lesions such as cavernous angioma, and FCD are more amenable to resection surgery with better outcome in lesional cases of extratemporal epilepsy than nonlesional cases [77].

In general, a complete resection of the epileptogenic brain region provides higher chances of seizure freedom but the risks of postoperative deficits would increase with the extension of resection. Therefore, the extent of resection should be weighed against such risks and individualized in every case.

While traditionally the mainstay of surgical treatment has been open brain resection, recent innovations have allowed for less invasive ablative techniques such as MRI-guided laser interstitial thermotherapy (LITT). A brief review of the different technical approaches and methods follows.

6.1. Surgery of Temporal Lobe Epilepsy

Anterior temporal lobectomy (ATL) is the most commonly performed surgery for the treatment of refractory temporal lobe epilepsy. Most patients who undergo ATL have concordant findings with MRI, VEEG, PET scan, and Neuropsychological testing. This procedure is performed using a variety of methods. The standard (en bloc) resection includes 3–6 cm of anterior temporal neocortex and 1–3 cm of mesial structures (amygdala and hippocampus) in the resection, with more limited resections performed on the language dominant side.

Selective amygdalo-hippocampectomy (SAH), is another modified surgical method that was developed to avoid the resection of lateral (neocortical) temporal tissue to minimize language deficits. The SAH is performed by accessing temporal horn through an incision in the middle temporal gyrus and selectively removing mesial temporal structures leaving the neocortical region intact [78]. Approximately 70% of properly selected surgical candidates become seizure-free following ATL, and the majority of remaining group benefit significantly by achieving seizure reduction and improved quality of life (QOL) [79,80].

6.2. Laser Ablation Surgery

Another method of operation used in temporal lobe as well as extra-temporal epilepsy is a less invasive procedure that can ablate epileptogenic focus that avoids craniotomy. Laser ablation surgery has been effectively applied in lesional and non-lesional cases including MTS, FCD, failed prior open surgery, and on deeper lesions that are inaccessible to open surgery.

This procedure has the advantage of selectively targeting small lesions responsible for seizures, with less pain and shorter hospitalization. The goal of an MRIgLITT (MRI-guided laser interstitial thermal therapy) system is to necrotize soft tissues through interstitial irradiation or thermal therapy under MRI guidance. Various pathologies have been treated with this technique including MTS, hypothalamic hamartoma, cavernous angioma, malformations of cortical development, and tuberous sclerotic lesions [81,82].

Mesial temporal epilepsy is also suitable for MRIgLITT. A history of prior laser ablative surgery would not preclude further open resection or repeat ablation procedure. Its safety and accuracy, in patients with difficult-to-localize seizures has been shown [83], and its efficacy has been reported as comparable to that of open surgery but with less morbidity [84,85] including in older patients [86]. Evidence suggests that selective surgical approaches for mesial temporal lobe epilepsy lower cognitive risks compared to the standard ATL. [87,88].

6.3. Corpus Callosotomy

Corpus collosotomy is a palliative surgical procedure devised to alleviate debilitating seizures in individuals with no focally resectable lesion [89]. Disconnection of the corpus callosum decreases the severity and frequency of rapidly propagating secondarily generalized focal seizures in patients who are not otherwise good surgical candidates particularly those with drop attacks (tonic or atonic seizures).

6.4. Multiple Subpial Transection Procedure

This technique introduced by Morrell in 1989 is primarily used for treatment of refractory epilepsies where resection is risky such as cases of seizure foci in the vicinity of, or with overlapping, eloquent cortex. The technique leaves the vertical columnar arrangement of the cortex intact thereby preserving function while preventing horizontal propagation of the seizure discharge. The procedure has been shown to be of value in cases of epileptogenic foci involving an eloquent cortex but less effective than excision surgery [90].

6.5. Hemispherectomy

Hemispherectomy (FH) is used mainly in children with refractory focal epilepsies confined to one hemisphere but also rarely in adults. Two main surgical techniques include anatomic hemispherectomy (which aims to resect all cortical tissue on one hemisphere along with various amounts of subcortical tissue), and functional hemispherectomy (which consist of more disconnection and less resection than the anatomic hemispherectomy.

Hemispherectomy is most commonly used in patients with perinatal strokes, Sturge-Weber syndrome, Rasmussen encephalitis, Infantile Hemiplegia Seizure Syndrome (IHSS), and hemimegalencephaly. Surgical candidates are selected based on a number of factors including severity of the underlying epilepsy, neurological condition, natural history of the underlying disease, and patient's age. Serious late post-op complications including superficial cerebral hemosiderosis and hydrocephalus can occur although modifications of the procedure have decreased these complications [91,92].

7. Non-Resective Surgical Treatments

Patients who are determined to be unqualified for seizure focus removal via resective surgery, such as those with multifocal seizures and those with overlapping seizure onset zone and eloquent cortex, may benefit from other surgical options. Attempts to control seizures through electrical stimulation started in 1960s have resulted in a few available treatment modalities starting with vagus nerve stimulation (VNS) in 1980s. Currently, there are two such procedures approved in the United States and one other approved elsewhere. These surgically implanted modalities do not require removal of brain tissue.

7.1. Vagus Nerve Stimulation (VNS)

This is an adjunctive therapy approved by the Food and Drug Administration (FDA) for refractory focal onset epilepsies with or without secondary generalization in patients 4 years or older. The VNS system is comprised of a battery generator programmed to deliver intermittent electrical stimuli via the left vagus nerve to the brain. The generator is implanted in the left upper chest and is tunneled under the skin to the vagus nerve. A retrospective review of efficacy of VNS in 48 patients with intractable partial epilepsy showed progressive decrease in mean seizure frequency by 26% after 1 year, 30% after 5 years, and 52% after 12 years [93].

7.2. Responsive Neurostimulation (RNS)

The RNS system was approved by the FDA in 2013 for medically refractory focal epilepsy. It provides cortical stimulation in response to ictal discharges detected by the RNS device. This is a programmable neurostimulator which is cranially implanted and connected to one or two depth and/or subdural cortical strip electrodes placed over the seizure foci. The Neurostimulator continually senses electrocorticographic activity and when detects onset of an ictal discharges, provides brief pulses of stimulation in response (closed-loop system).

The pivotal randomized study of 191 subjects showed a progressive reduction in the frequency of total disabling seizures in the treatment group (41.5%) compared to the sham group (9.4%) in the final months of the 2 year study [94].

7.3. Stimulation of the Anterior Nucleus of Thalamus (Deep Brain Stimulation, DBS)

In a multicenter trial, that evaluated stimulation of bilateral anterior nuclei of the thalamus, the treatment group had 29% greater reduction in seizures compared to the control group in the first month, and at least 50% seizure reduction in 54% of patients on 2 years follow up [95]. Stimulation of the anterior nucleus of thalamus has been approved as adjunctive treatment for drug-resistant focal epilepsy in adults, in the European Union since 2010, but its approval in the by FDA for use in the United States is still pending.

8. Postsurgical Outcome

Postsurgical outcome in epilepsy surgery is mainly assessed based on seizure-freedom or seizure reduction, neurocognitive functioning, and change in overall QOL. More detailed elements in measuring outcome include factors such as work eligibility, driving, etc. Of particular concern, are language and verbal memory function following temporal lobe surgeries.

Interestingly, despite the evidence supporting the efficacy and safety of epilepsy surgery, until 2001 there was no such evidence based on randomized, controlled trials (RCT). The first RCT included 80 patients with TLE randomly assigned to surgery (ATL), or continuing medical therapy, for one year to compare both seizure control outcome, and quality of life (QOL), disability, and mortality.

The study showed, for the first time, that after one year follow up, 58% of patients in the ATL group were seizure- free compared to only 8% in the medical therapy group ($p < 0.001$). Further, those in the surgical group had less frequent focal seizures and significantly better quality of life compared to the non-surgical group ($p < 0.001$ for both comparisons). In the surgical group 4 patients (10%) had mild language and memory deficits against one dead in the non-surgical group [1].

Since postsurgical outcome significantly depends on seizure type and seizure onset location, different types of epilepsies and surgical approaches are to be studied separately. A summary of a few selective outcome studies, including particular surgical methods used, follows.

8.1. Mesial Temporal Lobe Epilepsy with Hippocampal Sclerosis

8.1.1. Standard (en Block) Anterior Temporal Lobectomy (ATL)

Patients with MTLE and MTS can achieve long-term seizure control following standard ATL however the risk of seizure recurrence after ≥2 years exists. A review of 50 consecutive cases of ATL in patients with MTLE and MTS reported 82% seizure-freedom rate at 12 months, 76% at 24 months, and 64% at 63 months (mean follow up 5.8 years). The improvement in seizure control was associated with improved long-term QOL. The study also indicated reduction in AED as the major risk factor for seizure recurrence [96].

A more recent study of a similar group of patients undergoing standard ATL and followed for mean 6.7 years found 89% seizure-freedom, and significant improvement—defined as Engel Class I or II in 94%. Complete seizure freedom was seen in 103 patients (89%), and Engel Class I or II outcome (free of disabling seizures, or rare disabling seizures, respectively) in 94%. This study also determined concordance rates between the final resection site and the following factors: VEEG (100%), PET (100%), MRI (99.0%), and Wada test (90.4%). The lowest concordance was found with SPECT (84.6%), and neuropsychological testing (82.5%). Interestingly, a strong Wada memory lateralization was associated with excellent long-term seizure control, and vice versa, i.e., low disparity in scores between the sites predicted continued seizures [97].

8.1.2. Selective Amygdalohippocampectomy (SAH)

Patients with MTLE and MTS who undergo this technique may also achieve similar outcomes comparable to ATL with more extensive (neocortical temporal) resections [98]. While SAH was originally introduced as a method that could minimize neurocognitive deficits by preserving the neocortical temporal cortex, the controversy about its outcome involves neuropsychological deficits rather than seizure-freedom rates, as compared to standard ATL. While there is some evidence of SAH resulting in better cognitive function than ATL [99]. SAH performed in patients without adequate memory reserve can result in verbal memory deficits in dominant temporal lobe cases [100]. This topic has been reviewed by Schramm who concluded that there was considerable evidence of improved neuropsychological outcomes in SAH [101].

9. Neurocognitive Deficits Following ATL

At the presurgical baseline, patients with epilepsy, in particular those with TLE, often have some degree of cognitive impairment which may be further affected by surgery. It has been shown that larger temporal lobe resections provide better seizure control outcome however, they are more likely to induce cognitive deficits since larger resection risks including more of the functional tissue [102].

To further quantify association between the extent of temporal lobe resection and cognitive outcome a group of 47 right-handed patients with left MTLE who underwent ATL was studied. Cognitive changes were compared between limited resection (1–2 cm for mesial, and ≤4 cm for neocortical), and more extensive resection (>2 cm for mesial, and ≥4 cm for neocortical). There were no differences in cognitive outcome between the groups however, a negative correlation was found with age at seizure onset [103].

10. Conclusions

Drug-resistant epilepsy, defined as failure of two anti-seizure medications to completely control seizures, can be effectively treated with surgery. Therefore, proper use of diagnostic methods to determine a patient's eligibility for surgical treatment i.e., presurgical evaluation, is of utmost importance in managing patients with DRE. This process, as discussed in this article, also determines what type of surgical approach would be the safest and the most beneficial approach to the particular patient. With further developments in diagnostic and therapeutic methods, both presurgical evaluation and the surgical methods used, will improve constantly.

Supplementary Materials: The following are available online at http://www.mdpi.com/2076-3425/8/4/49/s1, Animated 3D reconstruction of subdural grid, strip, and depth electrodes.

Conflicts of Interest: The authors declare no conflicts of interest.

References

1. Wiebe, S.; Blume, W.T.; Girvin, J.P.; Eliasziw, M.; Effectiveness and Efficiency of Surgery for Temporal Lobe Epilepsy Study Group. A randomized, controlled trial of surgery for temporal lobe epilepsy. *N. Engl. J. Med.* **2001**, *345*, 311–318. [CrossRef] [PubMed]
2. Engel, J., Jr. Why is there still doubt to cut it out? *Epilepsy Curr.* **2013**, *13*, 198–204. [CrossRef] [PubMed]
3. Engel, J. Another Good Reason to Consider Surgical Treatment for Epilepsy More Often and Sooner. *Arch. Neurol.* **2011**, *68*, 707–708. [CrossRef] [PubMed]
4. Tellez-Zenteno, J.F.; Dhar, R.; Wiebe, S. Long-term seizure outcomes following epilepsy surgery: A systematic review and meta-analysis. *Brain* **2005**, *128*, 1188–1198. [CrossRef] [PubMed]
5. Scheffer, I.E.; Berkovic, S.; Capovilla, G.; Connolly, M.B.; Frence, J.; Guihoto, L.; Hirsch, E.; Jain, S.; Mathern, G.W.; Moshe, S.L.; et al. ILAE classification of the epilepsies: Position paper of the ILAE Commission for Classification and Terminology. *Epilepsia* **2017**, *58*, 512–521. [CrossRef] [PubMed]
6. Fisher, R.S.; Cross, J.H.; D'Souza, C.; French, J.A.; Haut, S.R.; Hiqurashi, N.; Hisch, E.; Jansen, F.E.; Laqae, L.; Moshe, S.L.; et al. Instruction manual for the ILAE 2017 operational classification of seizure types. *Epilepsia* **2017**, *58*, 531–542. [CrossRef] [PubMed]
7. Kwan, P.; Brodie, M.J. Early Identification of Refractory Epilepsy. *N. Engl. J. Med.* **2000**, *342*, 314–319. [CrossRef] [PubMed]
8. Kwan, P.; Arzimanoglou, A.; Berg, A.T.; Brodie, M.J.; Hauser, A.W.; Mathern, G.; Moshé, S.L.; Perucca, E.; Wiebe, S.; French, J. Definition of drug resistant epilepsy: Consensus proposal by the ad hoc Task Force of the ILAE Commission on Therapeutic Strategies. *Epilepsia* **2010**, *51*, 1069–1077. [CrossRef] [PubMed]
9. Lüders, H.O.; Najm, I.; Nair, D.; Widdess-Walsh, P.; Bingman, W. The epileptogenic zone: General principles. *Epileptic Disord.* **2006**, *8* (Suppl. 2), S1–S9. [PubMed]
10. Carreño, M.; Lüders, H.O. General Principles of Presurgical Evaluation. In *Epilepsy Surgery*, 2nd ed.; Lüders, H.O., Comair, Y.G., Eds.; Lippincott Williams & Wilkins: Philadelphia, PA, USA, 2001; pp. 185–200.
11. Spencer, S.S.; Sperling, M.R.; Shewmon, D.A. Intracranial electrodes. In *Epilepsy: A Comprehensive Textbook*; Engel, J., Pedley, T.A., Eds.; Lippincott-Raven: Philadelphia, PA, USA, 1997; pp. 1719–1747.
12. Cooper, R.; Winter, A.L.; Crow, H.J.; Grey Walter, W. Comparison of subcortical, cortical and scalp activity using chronically indwelling electrodes in man. *Electroencephalogr. Clin. Neurophysiol.* **1964**. [CrossRef]
13. Schulz, R.; Lüders, H.O.; Hoppe, M.; Tuxhorn, I.; May, T.; Ebner, A. Interictal EEG and ictal scalp EEG propagation are highly predictive of surgical outcome in mesial temporal lobe epilepsy. *Epilepsia* **2000**, *41*, 564–570. [CrossRef] [PubMed]
14. Pataraia, E.; Lurger, S.; Serles, W.; Lindinger, G.; Aull, S.; Leutmezer, F.; Bacher, J.; Olbrich, A.; Czech, T.; Novak, K.; et al. Ictal scalp EEG in unilateral mesial temporal lobe epilepsy. *Epilepsia* **1998**, *39*, 608–614. [CrossRef] [PubMed]
15. Kuzniecky, R.I. Neuroimaging of Epilepsy: Therapeutic Implications. *NeuroRx* **2005**, *2*, 384–393. [CrossRef] [PubMed]
16. Boxerman, J.L.; Hawash, K.; Bali, B.; Clarke, T.; Rogg, J.; Pal, D.K. Is Rolandic Epilepsy Associated with Abnormal Findings On Cranial MRI? *Epilepsy Res.* **2007**, *75*, 180–185. [CrossRef] [PubMed]
17. Spencer, D. MRI (Minimum Recommended Imaging) in Epilepsy. *Epilepsy Curr.* **2014**, *14*, 261–263. [CrossRef] [PubMed]
18. Bernasconi, A. Magnetic resonance imaging in intractable epilepsy: Focus on structural image analysis. In *Advances in Neurology*; Blume, W., Carlen, P.L., Wiebe, S., Young, A.B., Padjen, A., Starrveld, E., Eds.; Lippincott-Williams: Philadelphia, PA, USA, 2005; pp. 273–278.
19. Chassoux, F.; Landre, E.; Mellerio, C.; Turak, B.; Mann, M.W.; Daumas-Duport, C.; Chiron, C.; Devaux, B. Type II focal cortical dysplasia: Electroclinical phenotype and surgical outcome related to imaging. *Epilepsia* **2012**, *53*, 349–358. [CrossRef] [PubMed]
20. Adams, C.; Hwang, P.A.; Gilday, D.L.; Armstrong, D.C.; Becker, L.E.; Hoffman, H.J. Comparison of SPECT, EEG, CT, MRI, and pathology in partial epilepsy. *Pediatr. Neurol.* **1992**, *8*, 97–103. [CrossRef]

21. Devous, M.D., Sr.; Thisted, R.A.; Morgan, G.F.; Leroy, R.F.; Rowe, C.C. SPECT brain imaging in epilepsy: A meta-analysis. *J. Nucl. Med.* **1998**, *39*, 285–293. [PubMed]

22. Spencer, S.S. The relative contributions of MRI SPECT and PET imaging in epilepsy. *Epilepsia* **1994**, *35*, S72–S89. [CrossRef] [PubMed]

23. Weil, S.; Noachtar, S.; Arnold, S.; Yousry, T.A.; Winkler, P.A.; Tatsch, K. Ictal ECD-SPECT differentiates between temporal and extratemporal epilepsy: Confirmation by excellent postoperative seizure control. *Nucl. Med. Commun.* **2001**, *22*, 233–237. [CrossRef] [PubMed]

24. Ingvar, D.H. Epilepsy related to cerebral blood flow and metabolism. *Acta Psychiatr. Scand. Suppl.* **1984**, *313*, 21–26. [CrossRef] [PubMed]

25. Sunhee, K.; Mountz, J.M. Review Article: SPECT Imaging of Epilepsy: An Overview and Comparison with F-18 FDG PET. *Int. J. Mol. Imaging* **2011**, *2011*, 813028. [CrossRef]

26. Engel, J.J.; Brown, W.J.; Kuhl, D.E.; Phelps, M.E.; Mazziotta, J.C.; Crandall, P.H. Pathological findings underlying focal temporal lobe hypometabolism in partial epilepsy. *Ann. Neurol.* **1982**, *12*, 518–528. [CrossRef] [PubMed]

27. Meadow, K.J.; Loring, D.W. The Wada test for language and memory lateralization. *Neurology* **2005**, *65*, 659.

28. Willmann, O.; Wennberg, R.; May, T.; Woermann, F.G.; Pohlmann-Eden, B. The contribution of 18F-FDG PET in preoperative epilepsy surgery evaluation for patients with temporal lobe epilepsy A meta-analysis. *Seizure* **2007**, *16*, 509–520. [CrossRef] [PubMed]

29. Wheless, J.W.; Willmore, L.J.; Breier, J.I.; Kataki, M.; Smith, J.R.; King, D.W.; Meador, K.J.; Park, Y.D.; Loring, D.W.; Clifton, G.L.; et al. A comparison of magnetoencephalography, MRI and V-EEG in patients evaluated for epilepsy surgery. *Epilepsia* **1999**, *40*, 931–941. [CrossRef] [PubMed]

30. Heers, M.; Rampp, S.; Kaltenhäuser, M.; Pauli, E.; Rauch, C.; Dölken, M.T.; Stefan, H. Detection of epileptic spikes by magnetoencephalography and electroencephalography after sleep deprivation. *Seizure* **2010**, *19*, 397–403. [CrossRef] [PubMed]

31. Abou-Khalil, B. An update on determination of language dominance in screening for epilepsy surgery: The Wada test and newer noninvasive alternatives. *Epilepsia* **2007**, *48*, 442–455. [CrossRef] [PubMed]

32. Klöppel, S.; Büchel, C. Alternatives to the Wada test: A critical view of functional magnetic resonance imaging in preoperative use. *Curr. Opin. Neurol.* **2005**, *18*, 418–423. [CrossRef] [PubMed]

33. Binder, J.R.; Swanson, S.J.; Hammeke, T.A.; Morris, G.L.; Mueller, W.M.; Fischer, M.; Benbadis, S.; Frost, J.A.; Rao, S.M.; Hauqhton, V.M. Determination of language dominance using functional MRI: A comparison with the Wada test. *Neurology* **1996**, *46*, 978–984. [CrossRef] [PubMed]

34. Powell, H.W.; Duncan, J.S. Functional magnetic resonance imaging for assessment of language and memory in clinical practice. *Curr. Opin. Neurol.* **2005**, *18*, 161–166. [CrossRef] [PubMed]

35. Sabsevitz, D.S.; Swanson, S.J.; Hammeke, T.A.; Spanaki, M.V.; Possing, E.T.; Morris, G.L.; Mueller, W.M.; Binder, J.R. Use of preoperative functional neuroimaging to predict language deficits from epilepsy surgery. *Neurology* **2003**, *60*, 1788–1792. [CrossRef] [PubMed]

36. Binder, J.R.; Sabsevitz, D.S.; Swanson, S.J.; Hammeke, T.A.; Raghavan, M.; Mueller, W.M. Use of preoperative functional MRI to predict verbal memory decline after temporal lobe epilepsy surgery. *Epilepsia* **2008**, *49*, 1377–1394. [CrossRef] [PubMed]

37. Lehéricy, S.; Cohen, L.; Bazin, B.; Samson, S.; Giacomini, E.; Rougetet, R.; Hertz-Pannier, L.; Le Bihan, D.; Marsault, C.; Baulac, M. Functional MR evaluation of temporal and frontal language dominance compared with the Wada test. *Neurology* **2000**, *54*, 1625–1633. [CrossRef] [PubMed]

38. Dupont, S. Can functional MRI replace the Wada test? *Neurochirurgie* **2008**, *54*, 208–211. [CrossRef] [PubMed]

39. Constable, R.T.; Carpentier, A.; Pugh, K.; Westerveld, M.; Oszunar, Y.; Spencer, D.D. Investigation of the hippocampal formation using a randomized event-related paradigm and z-shimmed functional MRI. *Neuroimage* **2000**, *12*, 55–62. [CrossRef] [PubMed]

40. Binder, J.R. Functional MRI is a valid noninvasive alternative to Wada testing. *Epilepsy Behav.* **2011**, *20*, 214–222. [CrossRef] [PubMed]

41. Woermann, F.G.; Jokeit, H.; Luerding, R.; Freitag, H.; Schulz, R.; Guertler, S.; Okujava, M.; Wolf, P.; Tuxhorn, I.; Ebner, A. Language lateralization by Wada test and fMRI in 100 patients with epilepsy. *Neurology* **2003**, *61*, 699–701. [CrossRef] [PubMed]

42. Wada, J.A. A new method for determination of the side of cerebral dominance: A preliminary report on the intra-carotid injection of sodium amytal in man. *Igaku Seibutsugaku* **1949**, *14*, 221–222.

43. Simkins-Bullock, J. Beyond speech lateralization: A review of the variability, reliability, and validity of the intracarotid amobarbital procedure and its nonlanguage uses in epilepsy surgery candidates. *Neuropsychol. Rev.* **2000**, *10*, 41–74. [CrossRef] [PubMed]

44. Haag, A.; Knake, S.; Hamer, H.M.; Boesebeck, F.; Freitag, H.; Schulz, R.; Baum, P.; Helmstaedter, C.; Wellmer, J.; Urbach, H.; et al. The Wada test in Austrian, Dutch, German, and Swiss epilepsy centers from 2000 to 2005: A review of 1421 procedures. *Epilepsy Behav.* **2008**, *13*, 83–89. [CrossRef] [PubMed]

45. Bauer, P.R.; Reitsma, J.B.; Houweling, B.M.; Ferrier, C.H.; Ramsey, N.F. Can fMRI safely replace the Wada test for preoperative assessment of language lateralisation? A meta-analysis and systematic review. *J. Neurol. Neurosurg. Psychiatry* **2014**, *85*, 581–588. [CrossRef] [PubMed]

46. Dade, L.A.; Jones-Gotman, M. Sodium amobarbital memory tests: What do they predict? *Brain Cogn.* **1997**, *33*, 189–209. [CrossRef] [PubMed]

47. Loring, D.W.; Bowden, S.C.; Lee, G.P.; Meadow, K.J. Diagnostic utility of WADA memory Asymmetries: Sensitivity, specificity and likelihood ratio characterization. *Neuropsychology* **2009**, *23*, 687–693. [CrossRef] [PubMed]

48. Curot, J.; Denuelle, M.; Busigny, T.; Barragan-Jason, G.; Kany, M.; Tall, P.; Marlat, F.; Fabre, N.; Valton, L. Bilateral Wada test: Amobarbital or propofol? *Seizure* **2014**, *23*, 122–128. [CrossRef] [PubMed]

49. Möddel, G.; Lineweaver, T.; Schuele, S.U.; Reinholz, J.; Loddenkemper, T. Atypical language lateralization in epilepsy patients. *Epilepsia* **2009**, *50*, 1505–1516. [CrossRef] [PubMed]

50. Gilmore, R.L.; Heilman, K.M.; Schmidt, R.P.; Fennell, E.M.; Quisling, R. Anosognosia during Wada testing. *Neurology* **1992**, *42*, 925–927. [CrossRef] [PubMed]

51. Andelman, F.; Fried, I.; Neufeld, M.Y. Quality of life self-assessment as a function of lateralization of lesion in candidates for epilepsy surgery. *Epilepsia* **2001**, *42*, 549–555. [CrossRef] [PubMed]

52. Andelman, F.; Neufeld, M.Y.; Fried, L. Contribution of neuropsychology to epilepsy surgery. *Isr. J. Psychiatry Relat. Sci.* **2004**, *41*, 125–132. [PubMed]

53. Jones-Gotman, M.; Smith, M.L.; Zatorre, R.J. Neuropsychological testing for localizing and lateralizing the epileptogenic region. In *Surgical Treatment of the Epilepsies*, 2nd ed.; Engel, J., Jr., Ed.; Raven Press, Ltd.: New York, NY, USA, 1993; pp. 245–261.

54. Raspall, T.; Donate, M.; Boget, T.; Carreno, M.; Donaire, A.; Agudo, R.; Bargallo, N.; Rumia, J.; Setoain, X.; Pintor, L.; et al. Neuropsychological tests with lateralizing value in patients with temporal lobe epilepsy: Reconsidering material-specific theory. *Seizure* **2005**, *14*, 569–576. [CrossRef] [PubMed]

55. Squire, L.R.; Zola, S.M. Structure and function of declarative and non-declarative memory systems. *Proc. Natl. Acad. Sci. USA* **1996**, *93*, 13515–13522. [CrossRef] [PubMed]

56. Jenssen, S.; Liporace, J.; Nei, M.; O'Connor, M.J.; Sperling, M.R. Value of non-invasive testing when there are independent bitemporal seizures in the scalp EEG. *Epilepsy Res.* **2006**, *68*, 115–122. [CrossRef] [PubMed]

57. Spencer, S.S.; Nguyen, D.K.; Duckrow, R.B. Invasive EEG in presurgical evaluation of epilepsy. In *The Treatment of Epilepsy*, 3rd ed.; Shoruon, S., Perucca, E., Engel, J., Jr., Eds.; Wiley Blackwell: West Sussex, UK, 2009; pp. 767–798.

58. Schiller, Y.; Cascino, G.D.; Sharbrough, F.W. Chronic intracranial EEG monitoring for localizing the epileptogenic zone: An electroclinical correlation. *Epilepsia* **1998**, *39*, 1302–1308. [CrossRef] [PubMed]

59. Engel, J.; Van Ness, P.C.; Rasmussen, T.B.; Ojemann, L.M. Outcome with respect to epileptic seizures. In *Surgical Treatment of the Epilepsies*; Engel, J., Jr., Ed.; Raven Press: New York, NY, USA, 1993; pp. 609–621.

60. Fisher, R.S.; Webber, W.R.; Lesser, R.P.; Arroyo, S.; Uamatsu, S. High-frequency EEG activity at the start of seizures. *Clin. Neurophysiol.* **1992**, *9*, 441–448.

61. Gloor, P. Preoperative electroencephalographic investigation in temporal lobe epilepsy: Extracranial and intracranial recordings. *Can. J. Neurol. Sci.* **1991**, *18*, 554–558. [CrossRef] [PubMed]

62. Kaiboriboon, K.; Malkhachroum, A.M.; Zrik, A.; Daif, A.; Schiltz, N.M.; Labiner, D.M.; Lhatoo, S.D. Epilepsy surgery in the United States: Analysis of data from the national association of epilepsy centers. *Epilepsy Res.* **2015**, *116*, 105–109. [CrossRef] [PubMed]

63. Wong, C.H.; Birkett, J.; Byth, K.; Dexter, M.; Somerville, E.; Gill, D.; Chaseling, R.; Fearnside, M.; Bleasel, A. Risk factors for complications during intracranial electrode recording in presurgical evaluation of drug resistant partial epilepsy. *Acta Neurochir. (Wien)* **2009**, *151*, 37–50. [CrossRef] [PubMed]

64. Barnett, G.H.; Burgess, R.C.; Awad, I.A.; Skipper, G.J.; Edwards, C.R.; Luders, H. Epidural peg electrodes for the presurgical evaluation of intractable epilepsy. *Neurosurgery* **1990**, *27*, 113–115. [CrossRef] [PubMed]

65. Bancaud, J.; Dell, M.B. Technics and method of stereotaxic functional exploration of the brain structures in man (cortex, subcortex, central gray nuclei). *Rev. Neurol.* **1959**, *101*, 213–227. (In French) [PubMed]

66. Jeha, L.E.; Najm, I.; Bingaman, W.; Dinner, D.; Widdess-Walsh, P.; Luders, H. Surgical outcome and prognostic factors of frontal lobe epilepsy surgery. *Brain* **2007**, *130*, 574–584. [CrossRef] [PubMed]

67. Cossu, M.; Chabardes, S.; Hoffmann, D.; Lo Russo, G. Presurgical evaluation of intractable epilepsy using stereo-electro-encephalography methodology: Principles, technique and morbidity. *Neurochirurgie* **2008**, *54*, 367–373. (In French) [CrossRef] [PubMed]

68. Tanriverdi, T.; Ajlan, A.; Poulin, N.; Olivier, A. Morbidity in epilepsy surgery: An experience based on 2449 epilepsy surgery procedures from a single institution. *J. Neurosurg.* **2009**, *110*, 1111–1123. [CrossRef] [PubMed]

69. Guenot, M.; Isnard, J. Epilepsy and insula. *Neurochirurgie* **2008**, *54*, 374–381. [CrossRef] [PubMed]

70. Gonzalez-Martinez, J. The Stereo-Electroencephalography: The Epileptogenic Zone. *J. Clin. Neurophysiol.* **2016**, *33*, 522–529. [CrossRef] [PubMed]

71. Bulacio, J.C.; Chauvel, P.; McGonigal, A. Stereoelectroencephalography: Interpretation. *J. Clin. Neurophysiol.* **2016**, *33*, 503–510. [CrossRef] [PubMed]

72. Wu, S.; Kunhi Veedu, H.P.; Lhatoo, S.D.; Kouberssi, M.Z.; Miller, J.P.; Luders, H.O. Role of ictal baseline shifts and ictal high-frequency oscillations in stereo-electroencephalography analysis of mesial temporal lobe seizures. *Epilepsia* **2014**, *55*, 690–698. [CrossRef] [PubMed]

73. Chassoux, F.; Devaux, B.; Landré, E.; Turak, B.; Nataf, F.; Varlet, P.; Chodkiewicz, J.P.; Daumas-Duport, C. Stereoelectroencephalography in focal cortical dysplasia A 3D approach to delineating the dysplastic cortex. *Brain* **2000**, *123*, 1733–1751. [CrossRef] [PubMed]

74. Chabardès, S.; Kahane, P.; Minotti, L.; Tassi, L.; Grand, S.; Hoffmann, D.; Benabid, A.L. The temporopolar cortex plays a pivotal role in temporal lobe seizures. *Brain* **2005**, *128*, 1818–1831. [CrossRef] [PubMed]

75. Townsend, J.B.; Engel, J. Clinicopathological correlations of low voltage fast and high amplitude spike and wave medial temporal sterioencephalographic ictal onsets. *Epilepsia* **1991**, *32*, 21.

76. Motamedi, G.K.; Lesser, R.P.; Miglioretti, D.L.; Mizuno-Matsumoto, Y.; Gordon, B.; Webber, W.R.; Jackson, D.C.; Sepkuty, J.P.; Crone, N.E. Optimizing parameters for terminating cortical 1446 afterdischarges with pulse stimulation. *Epilepsia* **2002**, *43*, 836–846. [CrossRef] [PubMed]

77. Englot, D.J.; Wang, D.D.; Rolston, J.D.; Shih, T.T.; Chang, E.F. Rates and predictors of long-term seizure freedom after frontal lobe epilepsy surgery: A systematic review and meta-analysis. *J. Neurosurg.* **2012**, *116*, 1042–1048. [CrossRef] [PubMed]

78. Olivier, A.; Boling, W.W.; Tanriverdi, T. *Techniques in Epilepsy Surgery: The MNI Approach*; Cambridge University Press: Cambridge, UK, 2012.

79. Engel, J., Jr. Surgery for seizures. *N. Engl. J. Med.* **1996**, *334*, 647–657. [CrossRef] [PubMed]

80. Spencer, D.D.; Spencer, S.S.; Mattson, R.H.; Williamson, P.D.; Novelly, R.A. Access to the posterior medial temporal lobe structure in surgical treatment of temporal lobe epilepsy. *Neurosurgery* **1984**, *15*, 667–671. [CrossRef] [PubMed]

81. Gross, R.E.; Willie, J.T.; Drane, D.L. The Role of Stereotactic Laser Amygdalohippocampotomy in Mesial Temporal Lobe Epilepsy. *Neurosurg. Clin. N. Am.* **2016**, *27*, 37–50. [CrossRef] [PubMed]

82. Prince, E.; Hakimian, S.; Ko, A.L.; Ojemann, J.G.; Kim, M.S.; Miller, J.W. Laser Interstitial Thermal Therapy for Epilepsy. *Curr. Neurol. Neurosci. Rep.* **2017**, *17*, 63. [CrossRef] [PubMed]

83. Gonzalez-Martinez, J.; Vadera, S.; Mullin, J.; Enatsu, R.; Alexopoulos, A.V.; Patwardhan, R.; Bingaman, W.; Najm, I. Robot-assisted stereotactic laser ablation in medically intractable epilepsy: Operative technique. *Neurosurgery* **2014**, *10* (Suppl. 2), 167–172; [CrossRef] [PubMed]

84. Shukla, N.D.; Ho, A.L.; Pendharkar, A.V.; Sussman, E.S.; Halpern, C.H. Laser interstitial thermal therapy for the treatment of epilepsy: Evidence to date. *Neuropsychiatr. Dis. Treat.* **2017**, *13*, 2469–2475. [CrossRef] [PubMed]

85. Kang, J.Y.; Wu, C.; Tracy, J.; Lorenzo, M.; Evans, J.; Nei, M.; Skidmore, C.; Mintzer, S.; Sharan, A.D.; Sperling, M.R. Laser interstitial thermal therapy for medically intractable mesial temporal lobe epilepsy. *Epilepsia* **2016**, *57*, 325–334. [CrossRef] [PubMed]

86. Waseem, H.; Osborn, K.E.; Schoenberg, M.R.; Kelley, V.; Bozorg, A.; Cabello, D.; Benbadis, S.R.; Vale, F.L. Laser ablation therapy: An alternative treatment for medically resistant mesial temporal lobe epilepsy after age 50. *Epilepsy Behav.* **2015**, *51*, 152–157. [CrossRef] [PubMed]

87. Dredla, B.K.; Lucas, J.A.; Wharen, R.E.; Tatum, W.O. Neurocognitive outcome following stereotactic laser ablation in two patients with MRI-/PET+ mTLE. *Epilepsy Behav.* **2016**, *56*, 44–47. [CrossRef] [PubMed]

88. Gleissner, U.; Helmstaedter, C.; Schramm, J.; Elger, C.E. Memory outcome after selective amygdalohippocampectomy: A study in 140 patients with temporal lobe epilepsy. *Epilepsia* **2002**, *43*, 87–95. [CrossRef] [PubMed]

89. Stigsdotter-Broman, L.; Olsson, I.; Flink, R.; Rydenhag, B.; Malmgren, K. Long-term follow-up after callosotomy-a prospective, population-based, observational study. *Epilepsia* **2014**, *55*, 316–321. [CrossRef] [PubMed]

90. Morrell, F.; Whisler, W.W.; Bleck, T.P. Multiple subpial transection: A new approach to the surgical treatment of focal epilepsy. *J. Neurosurg.* **1989**, *70*, 231–239. [CrossRef] [PubMed]

91. Villemure, J.G.; Rasmussen, T. Functional hemispherectomy in children. *Neuropediatrics* **1993**, *24*, 53–55. [CrossRef] [PubMed]

92. Bien, C.G.; Schramm, J. Treatment of Rasmussen encephalitis half a century after its initial description: Promising prospects and a dilemma. *Epilepsy Res.* **2009**, *86*, 101–112. [CrossRef] [PubMed]

93. Uthman, B.M.; Reichl, A.M.; Dean, J.C.; Eisenschenk, S.; Gilmore, R.; Reid, S.; Roper, S.N.; Wilder, B.J. Effectiveness of vagus nerve stimulation in epilepsy patients: A 12-year observation. *Neurology* **2004**, *63*, 1124–1126. [CrossRef] [PubMed]

94. Heck, C.N.; King-Stephens, D.; Massey, A.D.; Nair, D.R.; Jobst, B.C.; Barkley, G.L.; Salanova, V.; Cole, A.J.; Smith, M.C.; Gwinn, R.P.; et al. Two-year seizure reduction in adults with medically intractable partial onset epilepsy treated with responsive neurostimulation: Final results of the RNS System Pivotal trial. *Epilepsia* **2014**, *55*, 432–441. [CrossRef] [PubMed]

95. Fisher, R.; Salanova, V.; Witt, T.; Worth, R.; Henry, T.; Gross, R.; Oommen, K.; Osori, I.; Nazzzaro, J.; Labar, D.; et al. Electrical stimulation of the anterior nucleus of thalamus for treatment of refractory epilepsy. *Epilepsia* **2010**, *51*, 899–908. [CrossRef] [PubMed]

96. Lowe, A.J.; David, E.; Kilpatrick, C.J.; Matkovic, Z.; Cook, M.J.; Kaye, A.; O'Brien, T.J. Epilepsy surgery for pathologically proven Hippocampal sclerosis provides long-term seizure control and improved quality of life. *Epilepsia* **2004**, *45*, 237–242. [CrossRef] [PubMed]

97. Elliott, R.E.; Bollo, R.J.; Berliner, J.L.; Silverberg, A.; Carlson, C.; Geller, E.B.; Barr, W.B.; Barr, W.B.; Devinsky, O.; Doyle, W.K. Anterior temporal lobectomy with amygdalohippocampectomy for mesial temporal sclerosis: Predictors of long-term seizure control. *J. Neurosurg.* **2013**, *119*, 261–272. [CrossRef] [PubMed]

98. Tanriverdi, T.; Olivier, A.; Poulin, N.; Andermann, F.; Dubeau, F. Long-term seizure outcome after mesial temporal lobe epilepsy surgery: Corticalamygdalohippocampectomy versus selective amygdalohippocampectomy. *J. Neurosurg.* **2008**, *108*, 517–524. [CrossRef] [PubMed]

99. Wieser, H.G.; Yaşargil, M.G. Selective amygdalohippocampectomy as a surgical treatment of mesiobasal limbic epilepsy. *Surg. Neurol.* **1982**, *17*, 445–457. [CrossRef]

100. Tanriverdi, T.; Dudley, R.W.; Hasan, A.; Al Jishi, A.; Al Hinai, Q.; Poulin, N.; Coulnat-Coulbois, S.; Olivier, A. Memory outcome after temporal lobe epilepsy surgery: Corticoamygdalohippocampectomy versus selective amygdalohippocampectomy. *J. Neurosurg.* **2010**, *113*, 1164–1175. [CrossRef] [PubMed]

101. Schramm, J. Temporal lobe epilepsy surgery and the quest for optimal extent of resection: A review. *Epilepsia* **2008**, *49*, 1296–1307. [CrossRef] [PubMed]

102. Helmstaedter, C. Cognitive outcomes of different surgical approaches in temporal lobe epilepsy. *Epileptic Disord.* **2013**, *15*, 221–239. [PubMed]

103. Wolf, R.L.; Ivnik, R.J.; Hirschorn, K.A.; Sharbrough, F.W.; Cascino, G.D.; Marsh, W.R. Neurocognitive efficiency following left temporal lobectomy: Standard versus limited resection. *J. Neurosurg.* **1993**, *79*, 76–83. [CrossRef] [PubMed]

brain
sciences

MDPI

Review

Quality of Life and Stigma in Epilepsy, Perspectives from Selected Regions of Asia and Sub-Saharan Africa

Warren Boling [1],*, Margaret Means [2] and Anita Fletcher [3]

[1]   Department of Neurosurgery, Loma Linda University, Loma Linda, CA 92354, USA
[2]   School of Medicine, University of Louisville, Louisville, KY 40202, USA; margaret.means@louisville.edu
[3]   Department of Neurology, University of Louisville, Louisville, KY 40202, USA; anita.fletcher@louisville.edu
*   Correspondence: wboling@llu.edu; Tel.: +1-909-558-4419

Received: 13 November 2017; Accepted: 29 March 2018; Published: 1 April 2018

Abstract: Epilepsy is an important and common worldwide public health problem that affects people of all ages. A significant number of individuals with epilepsy will be intractable to medication. These individuals experience an elevated mortality rate and negative psychosocial consequences of recurrent seizures. Surgery of epilepsy is highly effective to stop seizures in well-selected individuals, and seizure freedom is the most desirable result of epilepsy treatment due to the positive improvements in psychosocial function and the elimination of excess mortality associated with intractable epilepsy. Globally, there is inadequate data to fully assess epilepsy-related quality of life and stigma, although the preponderance of information we have points to a significant negative impact on people with epilepsy (PWE) and families of PWE. This review of the psychosocial impact of epilepsy focuses on regions of Asia and Sub-Saharan Africa that have been analyzed with population study approaches to determine the prevalence of epilepsy, treatment gaps, as well as factors impacting psychosocial function of PWE and their families. This review additionally identifies models of care for medically intractable epilepsy that have potential to significantly improve psychosocial function.

Keywords: medically intractable epilepsy; surgery of epilepsy; stigma; quality of life; developing world

1. Literature Review

1.1. Prevalence of Epilepsy

Epilepsy is an important public health problem representing 0.6% of the global burden of disease [1] that particularly impacts the people living in the lowest income countries where epilepsy incidence may be 10 fold more than in the developed world. The prevalence of epilepsy is about 0.8% in North America [2]. However, prevalence rates have been estimated to vary from 49 to 215 per 100,000 people among regions of Africa [3], and an overall prevalence of 15 in 1000 has been described for Sub-Saharan Africa (SSA) [4]. In Asia, median prevalence has been observed to be similar to that found in Europe and North America, although with wide variation and with significant prevalence differences between rural and urban dwellers [4].

Sub-Saharan Africa includes 11% of the world population, but benefits from only 1.3% of world income, and of the least wealthy countries in the world, 80% are in Africa [5]. With this wealth disparity as a backdrop, it is understandable that many Sub-Saharan residents experience inadequate access to epilepsy care. Additionally, Sub-Saharan Africa is a mostly rural dwelling population, and of the about 36% of Africans who live in cities, most live in extreme poverty situations. A resource poor environment combined with inadequate access to medical care may explain the very high epilepsy prevalence rates observed in Africa, confounding these challenges this difficult environment results in significant difficulty determining the true epidemiology of the disease [5]. However, several authors have undertaken population studies of epilepsy in isolated regions of Sub-Saharan Africa. In a cross-sectional study using a survey questionnaire of rural and urban dwelling population in Kampala, Uganda and surrounding countryside, Kaddumukasa, et al. identified an overall epilepsy prevalence of 13.3% [6]. Ae-Ngibise, et al. [7] conducted a cross-sectional survey of individuals residing in the middle part of Ghana. The authors determined that 10.1 per 1000 individuals residing in this West African region had active convulsive epilepsy. In rural Zambia a door-to-door survey was conducted by Birbeck, et al. to determine active epilepsy prevalence identified an adjusted prevalence of 12.5/1000 [8]. Age-specific prevalence rates were found to be highest among children 5–15 years old with a second smaller peak in the over 65-years old age group. Ngugi, et al. [9] undertook a population-based cross-sectional and case-control study of prevalence and risk factors of convulsive epilepsy in five African Health and Demographic Surveillance System centres (Kilifi, Kenya; Agincourt, South Africa; Iganga-Mayuge, Uganda; Ifakara, Tanzania; and Kintampo, Ghana) to evaluate the reasons for significant variation in estimates of prevalence in different African regions. Adjusted prevalence rates for convulsive epilepsy ranged from a high of 14.8 per 1000 in Tanzania to 7.0 per 1000 in South Africa. Several epilepsy-related risk factors were found to be associated with epilepsy prevalence. However, the epilepsy risk factor of most importance was exposure to parasites, and this risk factor likely explained most of the regional variation in epilepsy prevalence.

Regional variation of prevalence rates has been identified in Asia as well. In Southern provinces of China, a door-to-door survey identified the prevalence of epilepsy active within the last year and the last five years of 2.8% and 3.7%, respectively, and the prevalence of epilepsy was significantly higher in rural areas [10]. Wang, et al. also identified a contrast in epilepsy prevalence between rural and urban provinces with higher epilepsy rates in more rural provinces of Heilongjiang and Ningxia and relatively lower epilepsy rates in the more urban province of Henan [11]. Epidemiology studies of epilepsy in Pakistan and India have identified twice the epilepsy prevalence in rural compared with urban areas (6.23 versus 3.04 per 1000), and this difference was not related to neurocysticercosis infection rates [12,13].

The full epidemiological burden of epilepsy is likely underestimated in the door-to-door survey methodology most commonly employed in population studies of epilepsy in the developing world. In this method of case ascertainment in population-based studies of epilepsy, a questionnaire is administered to screen for members of a household with epilepsy. However, the commonly used screening questionnaire [14] is most sensitive to detect generalized tonic-clonic seizures so that individuals with partial complex, dyscognitive, or myoclonic seizures who do not experience generalized convulsions will be under-recognized. Furthermore, epilepsy associated stigma may lead to a reluctance to admit to an epileptic disorder in a household resulting in further underestimation of the number of prevalent cases.

A common finding worldwide is higher epilepsy rates in rural dwelling populations. The causes are most likely multifactorial and related to the recognized factors that result in a treatment gap in the developing world such as lower socioeconomic status of rural dwelling people and less access to medical clinics, health care providers, and medications. Additional possible factors impacting higher rates of epilepsy in rural areas include exposure to environmental etiologies of epilepsy such as parasites and less knowledge in rural areas of medical treatment options and greater reliance on traditional healers.

1.2. Epilepsy Related Mortality and Morbidity

In addition to an elevated prevalence, mortality of intractable epilepsy is high worldwide and particularly so in Sub-Saharan Africa. Epilepsy in East Africa has a substantially elevated standardized mortality ratio of 7.2 times age-matched controls, among the highest of the poorest areas of the world [15]. Low socioeconomic status is associated with an elevated risk of death from epilepsy [16,17], and epilepsy more frequently causes death in Sub-Saharan Africa than in developed countries [6]. The causes of epilepsy-related death include the underlying epilepsy etiologies, medical complications of epilepsy, trauma, suicide, and sudden unexplained death in epilepsy [18–20]. In Africa, burns are commonly encountered resulting from ubiquitous open-hearth fires in the home, and burn injuries are disproportionately encountered in people with epilepsy [21,22]. Birbeck reported burns or falls requiring hospitalization in 31% of epilepsy patients in rural Zambia [23]. In South Africa, 11% of accidental scalds presenting to one burn unit were epilepsy related [24]. In Tanzania, a long term follow-up study found that over 50% of deaths in epilepsy patients were related to status epilepticus, burns or drowning [25].

1.3. Epilepsy Treatment Gap

The treatment gap in epilepsy can be defined as the proportion of people with epilepsy who require treatment but do not receive it, and this parameter has been proposed as a useful concept to compare quality of care of epilepsy patients across regions and countries [26]. Treatment gap is strongly influenced by socioeconomics of a region or country. Meyer, et al. identified for every one-level decrease in World Bank income category, the treatment gap increased by a factor of 1.55 [26]. In the developing countries of Sub-Saharan Africa and Latin America, up to 90% of people with epilepsy have been reported to receive inadequate treatment or no treatment at all [12].

In Asian countries, the treatment gap is broadly estimated to be 29–98%, with a gap for most countries between 50% and 80% [12]. In Yueyang County, Hunan province, China, a more rural region of China, a door-to-door epidemiological survey of epilepsy was conducted by Pi, et al. [27] who found the lifetime prevalence rate of epilepsy was 4.5%. In this survey area, 35.0% of the people with epilepsy (PWE) had never been diagnosed or treated, 57.3% of PWE received non-standard treatments, and only 7.7% of patients received standard treatment. The non-standard treatments included the use of Chinese medicine in 42.7%. The authors identified in this rural Chinese region that 93.4% of patients with active epilepsy in the last year had a treatment gap as high as 96.6%. Hu, et al. [28] conducted questionnaire-based interviews in rural Western China to identify individuals with active convulsive epilepsy. The estimated prevalence of convulsive epilepsy was found to be 1.8 per 1000 in this population with a treatment gap estimated to be 66.3%. The authors were able to identify that a majority of individuals with epilepsy had consulted a doctor but failed to receive or adhere to an appropriate treatment program.

In Sub-Saharan Africa, the epilepsy treatment gap has not been adequately studied although the information we have points to the presence of a significant treatment gap. Koba Bora, et al. [29] identified in a charity run neuropsychiatric clinic in Lubumbashi, the second largest city in the Democratic Republic of Congo, the epilepsy treatment gap was above 67%. When asked to describe the cause of their epilepsy, 55.3% of patients or their families considered epilepsy to be of spiritual or religious origin. In a random cluster sampling survey, questionnaires were administered to individuals living in rural and urban parts of Rwanda to evaluate aspects of epilepsy in the country [30]. The investigators found a prevalence of 41/1000 of people having active epilepsy, and a treatment gap of 67.8% that consisted of 43% receiving no treatment and the remainder of the gap due to traditional healer treatment or a mixture of traditional and medical treatment. Mbuba, et al. conducted a cross-sectional survey and risk-factor analysis of the epilepsy treatment gap in Kilifi, Kenya [31]. PWE were identified in a cross-sectional survey to establish the prevalence of active convulsive epilepsy and to determine the presence of any treatment gap. In those with epilepsy, the investigators evaluated blood levels of antiepileptic drugs (AEDs) to corroborate the questionnaire results concerning medical treatment. The epilepsy treatment gap based on AEDs detected in blood samples was 62.4% and of the PWE who sought medical treatment, 77% were prescribed antiepileptic medications. The authors found that non-adherence to medical treatment was associated with negative beliefs and attitudes about epilepsy. South Africa is a unique country in Sub-Saharan Africa due to its economic size and high sophistication level of its institutions. The gross domestic product of South Africa is the third highest in Africa after Nigeria and Egypt. South Africa is generally considered to be the most advanced medical system in Africa regarding patient care delivery and training of healthcare professionals. Ngugi, et al. identified prevalence of active convulsant epilepsy in Sub-Saharan Africa to be lowest in Agincourt, South Africa and highest in Ifakara, Tanzania with a significant difference between the two [9].

Worldwide, particularly in low resource areas, a treatment gap is found commonly to be much higher in rural than in urban areas. Common themes are identified that contribute to a gap in treatment that includes poverty, a lack of access to AEDs, inadequate access to physicians trained to manage epilepsy, poor knowledge about epilepsy among the community and healthcare providers, failed models of healthcare delivery, as well as stigma of epilepsy arising largely from common misconceptions about epilepsy that it is contagious or has supernatural origins [23,32].

1.4. Stigma and Psychosocial Impact of Epilepsy

The burden of epilepsy in the developing world affects the individual with epilepsy, their family, and society in general. PWE are impacted by the physical dangers of a seizure that brings risk of injury and death. There is a psychosocial impact as well on PWE who may be stigmatized and marginalized in society due to epilepsy. A child with epilepsy in a low-resource family frequently cannot participate in the activities of family life such as collecting water for fear of drowning or cooking for fear of falling in the fire. Financial security is at risk due to the cost of epilepsy medication and treatment [33]. Additionally, some parents must leave their jobs and experience lost income in order to provide care for the epileptic child. Psychosocial development is affected by the stigma of epilepsy, which is present worldwide to some degree but is most visible in the developing world. In East Africa, epilepsy is frequently thought to result from demonic possession and a seizure is believed to be contagious. In Asia, misconceptions that epilepsy is hereditary is a common belief about the cause of epilepsy, which results in an inability to marry for the PWE and negatively impacts the entire family of PWE [34]. In both China and Vietnam, the lack of marriageability of PWE centers on two main issues: (1) the possibility that epilepsy is inheritable and so could be passed on to offspring, and (2) the perceived inability of PWE to carry out everyday living tasks and contribute adequately to the family economy [35]. The psychological burden of epilepsy in China was assessed by Wang, et al. [36] by asking PWE the question "What do you worry about most". Eighty percent of responders replied: "when the next seizure might occur" and the second most frequent response was: "facing discrimination". In rural areas, traditional beliefs shape the definitions and treatment of epilepsy, which results in patients and families seeking less Western medical treatment. In rural China, Wang et al. [11] found the attitudes towards PWE are mostly negative. About half of the population believed that PWE should not be employed, and epilepsy was often identified as a mental disorder equivalent to insanity. Even among highly educated individuals and community leaders in China, about half of people perceived epilepsy as a 'terrible' condition, and that the disease was not curable and may be hereditary [37]. Studies in China have identified high rates of stigma felt by PWE and their families [38]. In a survey of PWE in China, about 1/3 of Chinese with epilepsy thought they were treated differently by others because of their epilepsy and 50% chose to keep their epilepsy secret [39]. In both China and Vietnam, surveys of attitudes about PWE found most people thought PWE have low intelligence, negative changes in character, and are often unfit for school [35].

In East and South-central Africa, children with epilepsy often are prevented from attending school due to stigma. The epileptic child may be shunned from family meals. Stigma creates an environment that leads to few or no childhood friends or play [40]. Witchcraft and sorcery may be invoked to combat the suspected demonic origin of epilepsy, which has additional risks of injury for the PWE and additional costs from traditional healers often sought out and exhausted prior to medical solutions [41,42]. As a result, PWE frequently experience rejection and isolation due to commonly held misconceptions of epilepsy etiologies and transmissibility [43,44].

Many facets of life are impacted by epilepsy. Personal health security is threatened by epilepsy. Fear of injury, concern about the social consequences of having a seizure, and the stigmatization related to a diagnosis of epilepsy leads to social isolation. Baskind and Birbeck have described three forms of stigmatization associated with epilepsy: enacted, felt, and courtesy [45]. Enacted stigma occurs when the source of discrimination is another person. Felt stigma results from a fear of being discriminated against. Courtesy stigma occurs when someone close to a PWE, in relation or proximity, feels stigmatized. Studies have indicated that nearly half, and sometimes as many as 70%, of PWE report feeling stigma [46–49]. PWE who report greater felt stigma are more likely to suffer from low self-esteem, poorer psychological function, and more uncertainty about the future [50]. Psychiatric comorbidities, including depression and anxiety, are more prevalent in PWE who report greater felt stigma [51–54]. Importantly, quality of life is appreciably decreased in these individuals [53,54]. In fact, psychosocial factors related to the stigma of epilepsy have a greater impact on the quality of life of PWE than clinical variables, such as side effects of medications [54,55]. The elements contributing to felt stigma vary by region and culture, but those commonly cited include seizure worry, lack of social support, and seizure severity [56–59]. Family members and close supporters of PWE also report high levels of stigma [57,60]. Increased levels of felt and enacted stigma are associated with lower levels of education, lower socioeconomic status, minorities, and those with less exposure to PWE [53,61]. Although low socioeconomic status is highly correlated with felt stigma, low socioeconomic status alone does not account for felt stigma. Leaffer, et al. [53] found that quality of life, depressive symptoms, and social support have the greatest impact on reported felt stigma in PWE. The researchers also identified that felt stigma is significantly associated with quality of life in low socioeconomic status individuals and with depression severity and social support in individuals with high socioeconomic status. Stigma, in all forms, limits the personal, educational, and social opportunities of the person with epilepsy, leading to a significant impact on the quality of life of both the person with epilepsy and his or her family members.

The causes and consequences of epilepsy are heterogeneous across countries and regions, although there are significant knowledge gaps in many areas of the world including much of the Asian continent. It is clear that striking geographic differences exist in the etiologies of epilepsy [40–42]. There is as well cultural variability as to how a society relates to epilepsy [43–46]. The treatment gap found in low resource regions constrains the type and level of care that can be delivered to most of the people of the world with epilepsy since a majority of the worlds' population lives in low resource countries. There are formidable political and economic forces responsible for the worldwide treatment gap in epilepsy, yet regardless of one's ability to access appropriate treatment stigma continues to be an important factor in the life of PWE, and it is likely that stigma of epilepsy would need to be overcome in order to improve quality of life of individuals with epilepsy [9,19,47,48].

Projects in the developing world such as the Global Campaign against Epilepsy are currently working to change attitudes about epilepsy through education, to dispel myths, and reduce social isolation [62]. However, changes in attitude are very slow to develop. Evidence in the developed world points to seizure freedom as a critical factor to improving quality of life (QOL) in PWE, especially if achieved at an early age when social and cognitive skills are still developing [63,64]. The earlier in life seizures can be controlled the more likely it is that an individual will develop normal interpersonal skills and integrate into society (i.e., complete schooling, find work, and marry) [65]. Therefore, the most effective intervention to alleviate stigma is most likely to be elimination of the recurrent seizures.

1.5. Treatment of Medically Intractable Epilepsy

About 30% of PWE will fail medical treatment with ongoing and recurrent seizures despite medication, so-called intractable epilepsy. Intractable epilepsy has now been defined by the International League Against Epilepsy (ILAE) as recurring seizures that continue unabated despite trying and failing at least 2 anticonvulsant medications over at least one year [66]. Despite over 20 anticonvulsant medications available in North America and the European Union today for the treatment of epilepsy, the new medications are not more efficacious in controlling seizures compared to the old medications although newer medicines may have better side effect profiles. Intractable epilepsy, which is dangerous and life-threatening, significantly elevates mortality rates of PWE to 4.69 times that of age-matched controls [67]. The most effective means of reducing the morbidity and mortality of epilepsy is seizure control [68]. And intervention to treat medically intractable epilepsy with surgery to stop the seizures reduces mortality to that of the general population [67].

Surgery of epilepsy is a highly effective treatment to stop seizures for the majority of PWE [69,70]. The "low hanging fruit" of surgical treatment is surgery of temporal lobe epilepsy (TLE) due to the fact that TLE is common, frequently does not respond to medication, and has excellent opportunity to achieve seizure freedom from surgery. Boling et al. demonstrated in an epilepsy program established in Uganda, East Africa that surgery for medically intractable TLE can be accomplished in the developing world with good results on seizure freedom without serious complications [71]. In brief, the program recruited children with a history and seizure semiology typical for TLE from regional clinics of central and northern Uganda. Epilepsy characterization and seizure focus localization with video electroencephalography EEG, computed tomography CT brain imaging, and neuropsychology testing was done at CURE Childrens' Hospital of Uganda (CCHU). The hurdles that must be overcome in any developing world approach to epilepsy treatment are related primarily to inadequate expertise for reliable EEG video interpretation, a shortage of available technology, and a harsh environment. The lack of expertise was overcome with remote analysis of video EEG and CT imaging in North America by epilepsy experts and teleconference linkage with the developing world site. The array of sophisticated technology available to epilepsy programs in the developed world will not be available in the foreseeable future to the majority of people in the world, who live in countries that are severely resource constrained. Therefore, only technology reasonably available in the developing world site can feasibly be used in the establishment of a sustainable epilepsy program. At the time the program was established in Uganda, there was no MRI available in the country but CT was available, which provided adequate imaging identification of mesial temporal sclerosis in 50% of the surgical patients. Likewise, the candidates selected for surgery were straightforward TLE and most likely to benefit from surgery, so-called low hanging fruit. The harsh environment relates to frequent power outages, temperature extremes, and lack of technicians to repair and upkeep sophisticated and sensitive equipment. The CT scan functions well in this environment and it was learned that video EEG does as well. The details of the methods and results of the CCHU epilepsy program are available at Boling, et al. [71]. This program established a paradigm for sustainable epilepsy care in a setting of severe resource constraints that optimized available technology rather than maximizing technological requirements.

Fletcher, et al. then went on to analyze patients 8 years after an epilepsy evaluation at CCHU and found 70% were seizure-free after surgery, all the seizure-free patients had stopped their anti-epileptic medication, and none of the non-operated patients were seizure-free [72]. The authors retrospectively analyzed quality of life (QOL), stigma, and self-esteem of individuals both cured of epilepsy with surgery and non-operated with continued epilepsy. QOL was analyzed at long-term follow-up using an outcomes inventory developed and validated for people with epilepsy, the quality of life in epilepsy-31 inventory QOLIE-31 [73]. The test contains seven multi-item scales analyzing factors commonly impacting PWE and a single item that assesses overall health. The authors identified a significantly elevated QOL for patients in the surgical treatment group compared to non-surgical patients. Stigma was evaluated by Fletcher, et al. with a questionnaire developed by Joan Austin, et al. for children with epilepsy and their parents [74]. The Austin, et al. stigma measure has two scales, one for testing the child with epilepsy and another for the parent. The parent responds to five items on a 5-point Likert scale from 1 (strongly disagree) to 5 (strongly agree). To score, the five items are summed and divided by the number of items. A higher score reflects greater perceptions of stigma associated with their child having epilepsy. The child responds to 8 stigma related questions that ask how often they felt or acted in the ways described on a 5-point Likert scale from 1 (never) to 5 (very often). The questions broadly relate to subjective experiences with peer relationships, seizure and medication side effects, and day to day life challenges. To score, the items are summed and divided by the number of items. A higher score reflects greater perceptions of stigma. In this East African group of PWE, patients who were seizure-free after surgery showed significantly lower perceived stigma compared to non-surgical patients with continued epilepsy. The Austin parent survey likewise revealed lower perceived stigma in the parents with seizure-free children versus those parents with children who continued with seizures.

Fletcher, et al. [72] found significant psychosocial improvement in individuals who realized seizure freedom after surgery for intractable epilepsy compared with those who continued with epilepsy, and similar psychosocial benefits were identified in the parents/caretakers of seizure-free individuals. The authors also identified that the developing world model of TLE surgery was similarly effective as the developed world experience to achieve seizure freedom, surgery had very low risks, and seizure-free results were robust at long-term follow-up.

2. Conclusions

Sub-Saharan Africa appears to have a very high overall prevalence of epilepsy. However, there is significant regional variation Asia and Africa in regards to epilepsy prevalence that may be related to the presence of endemic disease and parasites that are epilepsy etiologies and/or related to the availability of appropriate medical treatment of epilepsy. In the developed and developing regions of the world, stigma appears to be an important factor that affects QOL in PWE. In most of the developing world, there are differences between rural and urban populations in regards to degree of stigma experienced by PWE, and the degree of epilepsy-related stigma is largely reduced in a population by a higher level of education and acceptance of medical models of epilepsy etiology and treatment.

Medically intractable epilepsy is a chronic, disabling and dangerous disease. However, surgery will stop seizures in the majority of well-selected individuals, and seizure freedom will reduce the elevated mortality risk of epilepsy to that of age-matched controls [75]. Additionally, surgical treatment of drug-resistant epilepsy in the developed world improves QOL and reduces stigma [76,77]. These outcomes, which underpin an emphasis in the developed world to surgically treat intractable epilepsy when a clear opportunity for seizure freedom exists, are translatable to the developing and low resource regions of the world. Surgery is the only opportunity to cure intractable epilepsy, and, therefore, should be considered as an important treatment approach in high and low resource regions of the world to eliminate excess mortality of intractable epilepsy and mitigate the psychosocial consequences that impact individuals with medically intractable epilepsy.

Acknowledgments: The authors received no grants or support for this research work, neither did the authors receive funds to cover the costs to publish in open access.

Author Contributions: Warren Boling contributed to the review of the literature, writing, and editing of this manuscript. Margaret Means contributed to the review of the literature, writing, and editing of this manuscript. Anita Fletcher contributed to the review of the literature, writing, and editing of this manuscript.

Conflicts of Interest: The authors report no conflicts of interest.

References

1. Murray, C.J.; Vos, T.; Lozano, R.; Naghavi, M.; Flaxman, A.D.; Michaud, C.; Ezzati, M.; Shibuya, K.; Salomon, J.A.; Abdalla, S.; et al. Disability-adjusted life years (DALYs) for 291 diseases and injuries in 21 regions, 1990–2010: A systematic analysis for the Global Burden of Disease Study. *Lancet* **2012**, *380*, 2197–2223. [CrossRef]
2. Hesdorffer, D.C.; Beck, V.; Begley, C.E.; Bishop, M.L.; Cushner-Weinstein, S.; Holmes, G.L.; Shafer, P.O.; Sirven, J.I.; Austin, J.K. Research implications of the Institute of Medicine Report, Epilepsy Across the Spectrum: Promoting Health and Understanding. *Epilepsia* **2013**, *54*, 207–216. [CrossRef] [PubMed]
3. Ngugi, A.K.; Kariuki, S.M.; Bottomley, C.; Kleinschmidt, I.; Sander, J.W.; Newton, C.R. Incidence of epilepsy: A systematic review and meta-analysis. *Neurology* **2011**, *77*, 1005–1012. [CrossRef] [PubMed]

4.   Yemadje, L.P.; Houinato, D.; Quet, F.; Druet-Cabanac, M.; Preux, P.M. Understanding the differences in prevalence of epilepsy in tropical regions. *Epilepsia* **2011**, *52*, 1376–1381. [CrossRef] [PubMed]

5.   Epilepsy in the WHO African Region—World Health Organization. Available online: http://www.who.int/mental_health/management/epilepsy_in_African-region.pdf (accessed on 26 December 2017).

6.   Kaddumukasa, M.; Mugeny, L.; Kaddumukasa, M.N.; Ddumba, E.; Devereaux, M.; Furlan, A.; Sajatovic, M.; Katabira, E. Prevalence and incidence of neurological disorders among adult Ugandans in rural and urban Mukono district; a cross-sectional study. *BMC Neurol.* **2016**, *16*, 227. [CrossRef] [PubMed]

7.   Ae-Ngibise, K.A.; Akpalu, B.; Ngugi, A.; Akpalu, A.; Agbokey, F.; Adjei, P.; Punguyire, D.; Bottomley, C.; Newton, C.; Owusu-Agyei, S. Prevalence and risk factors for Active Convulsive Epilepsy in Kintampo, Ghana. *Pan Afr. Med. J.* **2015**, *21*, 29. [CrossRef] [PubMed]

8.   Birbeck, G.L.; Kalichi, E.M. Epilepsy prevalence in rural Zambia: A door-to-door survey. *Trop. Med. Int. Health* **2004**, *9*, 92–95. [CrossRef] [PubMed]

9.   Ngugi, A.K.; Bottomley, C.; Kleinschmidt, I.; Wagner, R.G.; Kakooza-Mwesige, A.; Ae-Ngibise, K.; Owusu-Agyei, S.; Masanja, H.; Kamuyu, G.; SEEDS Group; et al. Prevalence of active convulsive epilepsy in Sub-Saharan Africa and associated risk factors: Cross-sectional and case-control studies. *Lancet Neurol.* **2013**, *12*, 253–263. [CrossRef]

10. Pi, X.; Zhou, L.; Cui, L.; Liu, A.; Zhang, J.; Ma, Y.; Liu, B.; Cai, C.; Zhu, C.; Zhou, T.; et al. Prevalence and clinical characteristics of active epilepsy in southern Han Chinese. *Seizure-Eur. J. Epilepsy* **2014**, *23*, 636–640. [CrossRef] [PubMed]

11. Wang, W.; Wu, J.; Dai, X.; Ma, G.; Yang, B.; Wang, T.; Yuan, C.; Ding, D.; Hong, Z.; Kwan, P.; et al. Global campaign against epilepsy: Assessment of a demonstration project in rural China. *Bull. WHO* **2008**, *86*, 964–969. [PubMed]

12. Mac, T.L.; Tran, D.; Quet, F.; Odermatt, P.; Preux, P.; Tan, C.T. Epidemiology, aetiology, and clinical management of epilepsy in Asia: A systematic review. *Lancet Neurol.* **2007**, *6*, 533–543. [CrossRef]

13. Rajshekhar, V.; Raghava, M.V.; Prabhakaran, V.; Oommen, A.; Muliyil, J. Active epilepsy as an index of burden of neurocysticercosis in Vellore district, India. *Neurology* **2006**, *67*, 2135–2139. [CrossRef] [PubMed]

14. Placencia, M.; Sander, J.W.; Shorvon, S.D.; Ellison, R.H.; Cascante, S.M. Validation of a screening questionnaire for the detection of epileptic seizures in epidemiological studies. *Brain* **1992**, *115*, 783–794. [CrossRef] [PubMed]

15. Newton, C.R.; Garcia, H.H. Epilepsy in poor regions of the world. *Lancet* **2012**, *380*, 1193–1201. [CrossRef]

16. Nayel, M.H. Mutual benefits from epilepsy surgery in developed and developing countries. *Epilepsia* **2000**, *41* (Suppl. 4), S28–S30. [CrossRef] [PubMed]

17. Cockerell, O.C.; Johnson, A.L.; Sander, J.W.A.S.; Hart, Y.M.; Goodridge, D.M.G.; Shorvon, S.D. Mortality from epilepsy: Results from a prospective population-based study. *Lancet* **1994**, *344*, 918–921. [CrossRef]

18. Gaitatzis, A.; Sander, J.W. The mortality of epilepsy revisited. *Epileptic Disord.* **2004**, *6*, 3–13. [PubMed]

19. Pompili, M.; Girardi, P.; Ruberto, A.; Tatarelli, R. Suicide in the epilepsies: A meta-analytic investigation of 29 cohorts. *Epilepsy Behav.* **2005**, *7*, 305–310. [CrossRef] [PubMed]

20. Tellez-Zenteno, J.F.; Ronquillo, L.H.; Wiebe, S. Sudden unexpected death in epilepsy: Evidence-based analysis of incidence and risk factors. *Epilepsy Res.* **2005**, *65*, 101–115. [CrossRef] [PubMed]

21. Birbeck, G.L.; Munsat, T. Neurologic services in Sub-Saharan Africa: A case study among Zambian primary healthcare workers. *J. Neurol. Sci.* **2002**, *200*, 75–78. [CrossRef]

22. Amayo, E.O. Kenya. *Pract. Neurol.* **2006**, *6*, 261. [CrossRef]

23. Birbeck, G.L. Seizures in rural Zambia. *Epilepsia* **2000**, *41*, 277–281. [CrossRef] [PubMed]

24. Hudson, D.A.; Duminy, F. Hot water burns in Cape Town. *Burns* **1995**, *21*, 54–56. [CrossRef]

25. Jilek-Aall, L.; Rwiza, H.T. Prognosis of epilepsy in a rural African community: A 30-year follow-up of 164 patients in an outpatient clinic in rural Tanzania. *Epilepsia* **1992**, *33*, 645–650. [CrossRef] [PubMed]

26. Bulletin of the World Health Organization. Available online: http://www.who.int/bulletin/volumes/88/4/09-064147/en/ (accessed on 26 December 2017).

27. Pi, X.; Cui, L.; Liu, A.; Zhang, J.; Ma, Y.; Liu, B.; Cai, C.; Zhu, C.; Zhou, T.; Chen, J.; et al. Investigation of prevalence, clinical characteristics and management of epilepsy in Yueyang city of China by door-to-door survey. *Epilepsy Res.* **2012**, *101*, 129–134. [CrossRef] [PubMed]

28. Hu, J.; Si, Y.; Zhou, D.; Mu, J.; Li, J.; Liu, L.; Zhu, C.R.; Deng, Y.; He, J.; Zhang, N.M.; et al. Prevalence and treatment gap of active convulsive epilepsy: A large community-based survey in rural West China. *Seizure-Eur. J. Epilepsy* **2014**, *23*, 333–337. [CrossRef] [PubMed]

29. Koba Bora, B.; Lez, D.M.; Luwa, D.O.; Baguma, M.B.; Katumbay, D.T.; Kalula, T.K.; Mesu'a Kabwa, P.L. Living with epilepsy in Lubumbashi (Democratic Republic of Congo): Epidemiology, risk factors and treatment gap. *Pan Afr. Med. J.* **2015**, *26*, 303. [CrossRef] [PubMed]

30. Sebera, F.; Munyandamutsa, N.; Teuwen, D.E.; Ndiaye, I.P.; Diop, A.G.; Tofighy, A.; Boon, P.; Dedeken, P. Addressing the treatment gap and societal impact of epilepsy in Rwanda–Results of a survey conducted in 2005 and subsequent actions. *Epilepsy Behav.* **2015**, *46*, 126–132. [CrossRef] [PubMed]

31. Mbuba, C.K.; Ngugi, A.K.; Fegan, G.; Ibinda, F.; Muchohi, S.N.; Nyundo, C.; Odhiambo, R.; Edwards, T.; Odermatt, P.; Carter, J.A.; et al. Risk factors associated with the epilepsy treatment gap in Kilifi, Kenya: A cross-sectional study. *Lancet Neurol.* **2012**, *11*, 688–696. [CrossRef]

32. Reis, R. Evil in the body, disorder of the brain: Interpretation of epilepsy and the treatment gap in Swaziland. *Trop. Geogr. Med.* **1994**, *46*, S40–S43. [PubMed]

33. Nsengiyumva, G.; Druet-Cabanac, M.; Nzisabira, L.; Preux, P.M.; Vergnenègre, A. Economic evaluation of epilepsy in Kiremba (Burundi): A case-control study. *Epilepsia* **2004**, *45*, 673–677. [CrossRef] [PubMed]

34. Snape, D.; Wang, W.; Wu, J.; Jacoby, A.; Baker, G.A. Knowledge gaps and uncertainties about epilepsy: Findings from an ethnographic study in China. *Epilepsy Behav.* **2009**, *14*, 172–178. [CrossRef] [PubMed]

35. Jacoby, A.; Wang, W.; Dang Vu, T.D.; Wu, J.; Snape, D.; Aydemir, N.; Parr, J.; Reis, R.; Begley, C.; de Boer, H.; et al. Meanings of epilepsy in its sociocultural context and implications for stigma: Findings from ethnographic studies in local communities in China and Vietnam. *Epilepsy Behav.* **2008**, *12*, 286–297. [CrossRef] [PubMed]

36. Wang, W.; Zhao, D.; Wu, J.; Wang, T.; Dai, X.; Ma, G.; Yang, B.; Yuan, C.; Bell, G.S.; de Boer, H.M.; et al. Changes in knowledge, attitude, and practice of people with epilepsy and their families after an intervention in rural China. *Epilepsy Behav.* **2009**, *16*, 76–79. [CrossRef] [PubMed]

37. Yang, R.; Wang, W.; Snape, D.; Chen, G.; Zhang, L.; Wu, J.Z.; Baker, G.A.; Zheng, X.Y.; Jacoby, A. Stigma of People with Epilepsy in China: Views of health professionals, teachers, employers and community leaders. *Epilepsy Behav.* **2011**, *21*, 261–266. [CrossRef] [PubMed]

38. Kleinman, A.; Wang, W.Z.; Li, S.C.; Cheng, X.M.; Dai, X.Y.; Li, K.T.; Kleinman, J. The social course of epilepsy: Chronic illness as social experience in interior China. *Soc. Sci. Med.* **1995**, *40*, 1319–1330. [CrossRef]

39. Li, S.; Wu, J.; Wang, W.; Jacoby, A.; de Boer, H.; Sander, J.W. Stigma and epilepsy: The Chinese perspective. *Epilepsy Behav.* **2010**, *17*, 242–245. [CrossRef] [PubMed]

40. Matuja, W.B.; Rwiza, H.T. Knowledge, attitude and practice (KAP) towards epilepsy in secondary school students in Tanzania. *Cent. Afr. J. Med.* **1994**, *40*, 13–18. [PubMed]

41. Osuntokun, B.O. Epilepsy in the developing countries. The Nigerian profile. *Epilepsia* **1972**, *13*, 107–111. [CrossRef] [PubMed]

42. Osuntokun, B.O. Epilepsy in Africa. Epidemiology of epilepsy in developing countries in Africa. *Trop. Geogr. Med.* **1978**, *30*, 23–32. [PubMed]

43. Baker, G.A. The psychosocial burden of epilepsy. *Epilepsia* **2002**, *43* (Suppl. 6), 26–30. [CrossRef] [PubMed]

44. Jilek-Aall, L.; Jilek, M.; Kaaya, J.; Mkombachepa, L.; Hillary, K. Psychosocial study of epilepsy in Africa. *Soc. Sci. Med.* **1997**, *45*, 783–795. [CrossRef]

45. Baskind, R.; Birbeck, G.L. Epilepsy-associated stigma in Sub-Saharan Africa: The social landscape of a disease. *Epilepsy Behav.* **2005**, *7*, 68–73. [CrossRef] [PubMed]

46. Sleeth, C.; Drake, K.; Labiner, D.M.; Chong, J. Felt and enacted stigma in elderly persons with epilepsy: A qualitative approach. *Epilepsy Behav.* **2016**, *55*, 108–112. [CrossRef] [PubMed]

47. Lee, G.H.; Lee, S.A.; No, S.K.; Lee, S.M.; Ryu, J.Y.; Jo, K.D.; Kwon, J.H.; Kim, O.J.; Park, H.; Kwon, O.Y.; et al. Factors contributing to the development of perceived stigma in people with newly diagnosed epilepsy: A one-year longitudinal study. *Epilepsy Behav.* **2016**, *54*, 1–6. [CrossRef] [PubMed]

48. Baker, G.; Brooks, J.; Buck, D.; Jacoby, A. The Stigma of Epilepsy: A European Perspective. *Epilepsia* **1999**, *41*, 98–104. [CrossRef]

49. Luna, J.; Nizard, M.; Becker, D.; Gerard, D.; Cruz, A.; Ratsimbazafy, V.; Dumas, M.; Cruz, M.; Preux, P.M. Epilepsy-associated levels of perceived stigma, their associations with treatment, and related factors: A cross-sectional study in urban and rural areas in Ecuador. *Epilepsy Behav.* **2017**, *68*, 71–77. [CrossRef] [PubMed]

50. Jacoby, A.A. Felt versus enacted stigma: A concept revisited. Evidence from a study of people with epilepsy in remission. *Soc. Sci. Med.* **1994**, *38*, 269–274. [CrossRef]

51. Fernandez, P.; Snape, D.; Beran, R.; Jacoby, A. Epilepsy stigma: What do we know and where next? *Epilepsy Behav.* **2011**, *22*, 55–62. [CrossRef] [PubMed]

52. Wang, Y.H.; Haslam, M.; Yu, M.; Ding, J.; Lu, Q.; Pan, F. Family functioning, marital quality and social support in Chinese patients with epilepsy. *Health Qual. Life Outcomes* **2015**, *13*, 10. [CrossRef] [PubMed]

53. Leaffer, E.; Hesdorffer, D.; Begley, C. Psychosocial and sociodemographic associates of felt stigma in epilepsy. *Epilepsy Behav.* **2014**, *37*, 104–109. [CrossRef] [PubMed]

54. Suurmeijer, T.P.B.M.; Reuvekamp, M.F.; Aldenkamp, B.P. Social Functioning, Psychological Functioning, and Quality of Life in Epilepsy. *Epilepsia* **2001**, *42*, 1160–1168. [CrossRef] [PubMed]

55. Hermann, B.; Whitman, S.; Wyler, A.; Anton, M.; Vanderzwagg, R. Psychosocial Predictors of Psychopathology in Epilepsy. *Br. J. Psychiatry* **1990**, *156*, 98–105. [CrossRef]

56. Austin, J.K.; Perkins, S.M.; Dunn, D.W. A model for internalized stigma in children and adolescents with epilepsy. *Epilepsy Behav.* **2014**, *36*, 74–79. [CrossRef] [PubMed]

57. Benson, A.; O'Toole, S.; Lambert, V.; Gallagher, P.; Shahwan, A.; Austin, J.K. The stigma experiences and perceptions of families living with epilepsy: Implications for epilepsy-related communication within and external to the family unit. *Patient Educ. Couns.* **2016**, *99*, 1473–1481. [CrossRef] [PubMed]

58. Smith, G.; Ferguson, P.L.; Saunders, L.L.; Wagner, J.L.; Wannamaker, B.B.; Selassie, A.W. Psychosocial factors associated with stigma in adults with epilepsy. *Epilepsy Behav.* **2009**, *16*, 484–490. [CrossRef] [PubMed]

59. Kanemura, H.; Sano, F.; Ohyama, T.; Sugita, K.; Aihara, M. Seizure severity in children with epilepsy is associated with their parents' perception of stigma. *Epilepsy Behav.* **2016**, *63*, 42–45. [CrossRef] [PubMed]

60. Rood, J.E.; Schultz, J.R.; Rausch, J.R.; Modi, A.C. Examining perceived stigma of children with newly-diagnosed epilepsy and their caregivers over a two-year period. *Epilepsy Behav.* **2014**, *39*, 38–41. [CrossRef] [PubMed]

61. Herrmann, L.K.; Welter, E.; Berg, A.T.; Perzynski, A.T.; Van Doren, J.R.; Sajatovic, M. Epilepsy misconceptions and stigma reduction: Current status in Western countries. *Epilepsy Behav.* **2016**, *60*, 165–173. [CrossRef] [PubMed]

62. World Health Organization. Global Campaign against Epilepsy: Out of the Shadows. Available online: http://www.who.int/mental_health/management/globalepilepsycampaign/en/index.html (accessed on 11 January 2008).

63. McLachlan, R.S.; Rose, K.J.; Derry, P.A.; Bonnar, C.; Blume, W.T.; Girvin, J.P. Health-related quality of life and seizure control in temporal lobe epilepsy. *Ann. Neurol.* **1997**, *41*, 482–489. [CrossRef] [PubMed]

64. Gilliam, F.; Kuzniecky, R.; Meador, K.; Martin, R.; Sawrie, S.; Viikinsalo, M.; Morawetz, R.; Faught, E. Patient-oriented outcome assessment after temporal lobectomy for refractory epilepsy. *Neurology* **1999**, *53*, 687–694. [CrossRef] [PubMed]

65. Rausch, R.; Crandall, P.H. Psychological status related to surgical control of temporal lobe seizures. *Epilepsia* **1982**, *23*, 191–202. [CrossRef] [PubMed]

66. Kwan, P.; Arzimanoglou, A.; Berg, A.T.; Brodie, M.J.; Allen Hauser, W.; Mathern, G.; Moshé, S.L.; Perucca, E.; Wiebe, S.; French, J. Definition of drug resistant epilepsy: Consensus proposal by the ad hoc Task Force of the ILAE Commission on Therapeutic Strategies. *Epilepsia* **2010**, *51*, 1069–1077. [CrossRef] [PubMed]

67. Sperling, M.R.; Feldman, H.; Kinman, J.; Liporace, J.D.; O'Connor, M.J. Seizure control and mortality in epilepsy. *Ann. Neurol.* **1999**, *46*, 45–50. [CrossRef]

68. Salanova, V.; Markand, O.; Worth, R. Temporal lobe epilepsy surgery: Outcome, complications, and late mortality rate in 215 patients. *Epilepsia* **2002**, *43*, 170–174. [CrossRef] [PubMed]

69. Wiebe, S.; Blume, W.T.; Girvin, J.P.; Eliasziw, M. A randomized, controlled trial of surgery for temporal-lobe epilepsy. *N. Engl. J. Med.* **2001**, *345*, 311–318. [CrossRef] [PubMed]

70. Alonso-Vanegas, M.A.; Freire Carlier, I.D.; San-Juan, D.; Martínez NPsych, A.R.; Trenado, C. Parahippocampectomy as a new surgical approach to mesial temporal lobe epilepsy due to hippocampal sclerosis: A pilot randomized comparative clinical trial. *World Neurosurg.* **2017**. [CrossRef]

71. Boling, W.; Palade, A.; Wabulya, A.; Longoni, N.; Warf, B.; Nestor, S.; Alpitsis, R.; Bittar, R.; Howard, C.; Andermann, F. Surgery for Pharmacoresistant Epilepsy in the Developing World: A Pilot Study. *Epilepsia* **2009**, *50*, 1256–1261. [CrossRef] [PubMed]
72. Fletcher, A.; Sims-Williams, H.; Wabulya, A.; Boling, W. Stigma and quality of life at long-term follow-up after surgery for epilepsy in Uganda. *Epilepsy Behav.* **2015**, *52*, 128–131. [CrossRef] [PubMed]
73. Cramer, J.A.; Perrine, K.; Devinsky, O.; Bryant-Comstock, L.; Meador, K.; Hermann, B. Development and cross-cultural translation of a 31-item quality of life questionnaire (QOLIE-31). *Epilepsia* **1998**, *39*, 81–88. [CrossRef] [PubMed]
74. Austin, J.K.; MacLeod, J.; Dunn, D.W.; Shen, J.; Perkins, S.M. Measuring stigma in children with epilepsy and their parents: Instrument development and testing. *Epilepsy Behav.* **2004**, *5*, 472–482. [CrossRef] [PubMed]
75. Sperling, M.R.; O'Connor, M.J.; Saykin, A.J.; Plummer, C. Temporal lobectomy for refractory epilepsy. *JAMA* **1996**, *276*, 470–475. [CrossRef] [PubMed]
76. Markand, O.; Salanova, V.; Whelihan, E.; Emsley, C.L. Health-related quality of life outcome in medically refractory epilepsy treated with anterior temporal lobectomy. *Epilepsia* **2000**, *41*, 749–759. [CrossRef] [PubMed]
77. Jobst, B.C.; Cascino, G.D. Resective epilepsy surgery for drug-resistant focal epilepsy: A review. *JAMA* **2015**, *313*, 285–923. [CrossRef] [PubMed]

**brain
sciences**

MDPI

Article

Epilepsy and Neuromodulation—Randomized Controlled Trials

Churl-Su Kwon [1,2,3,*], **Valeria Ripa** [4], **Omar Al-Awar** [5], **Fedor Panov** [3], **Saadi Ghatan** [3] **and Nathalie Jetté** [1,2]

[1] Department of Neurology, Icahn School of Medicine at Mount Sinai, New York, NY 10029, USA; nathalie.jette@mssm.edu
[2] Department of Population Health Science and Policy, Icahn School of Medicine at Mount Sinai, New York, NY 10029, USA
[3] Department of Neurosurgery, Icahn School of Medicine at Mount Sinai, New York, NY 10029, USA; fedor.panov@mountsinai.org (F.P.); Saadi.Ghatan@mountsinai.org (S.G.)
[4] St. George's Medical School, St. George, Grenada; valeriaripa87@gmail.com
[5] Department of Neurosurgery, Oxford University, John Radcliffe Hospital, Oxford OX3 9DU, UK; awar_omar@hotmail.com
* Correspondence: churlsu.kwon@mssm.edu; Tel.: +1-212-241-9951; Fax: +1-646-537-9515

Received: 19 March 2018; Accepted: 16 April 2018; Published: 18 April 2018

Abstract: Neuromodulation is a treatment strategy that is increasingly being utilized in those suffering from drug-resistant epilepsy who are not appropriate for resective surgery. The number of double-blinded RCTs demonstrating the efficacy of neurostimulation in persons with epilepsy is increasing. Although reductions in seizure frequency is common in these trials, obtaining seizure freedom is rare. Invasive neuromodulation procedures (DBS, VNS, and RNS) have been approved as therapeutic measures. However, further investigations are necessary to delineate effective targeting, minimize side effects that are related to chronic implantation and to improve the cost effectiveness of these devices. The RCTs of non-invasive modes of neuromodulation whilst showing much promise (tDCS, eTNS, rTMS), require larger powered studies as well as studies that focus at better targeting techniques. We provide a review of double-blinded randomized clinical trials that have been conducted for neuromodulation in epilepsy.

Keywords: epilepsy; neuromodulation; randomized clinical trials (RCT); deep brain stimulation (DBS); transcranial direct current stimulation (tDCS); vagal nerve stimulation (VNS); external trigeminal nerve stimulation (eTNS); repetitive transcranial magnetic stimulation (rTMS); responsive neurostimulation (RNS)

1. History of Neuromodulation

There has been a long experimental history of cortical and deep brain neuromodulation in epilepsy. Only in the last 20 years however, through improved knowledge of brain networks, accurate stereotactic neurosurgery and robust trial design has this been translated into accepted clinical use. The dawn of human stereotactic deep brain stimulation (DBS) can be seen in medial thalamotomy psychosurgery by Spiegel and Wycis in 1947 in an attempt to decrease the brutal but commonly performed frontal lobotomies [1]. Gildenberg, a fellow at the time to Spiegel and Wycis, stated that intraoperative brain stimulation was used as a means of investigating the target prior to lesioning. Hence, from the origins of DBS, electrical stimulation was adopted as a physiological tool to evaluate deep brain structures [2]. Epilepsy too became an interest amongst early DBS pioneers, and initial studies looked to target epileptic foci with implantation of chronic stereotactic deep brain electrodes for interrogation and intermittent stimulation [3]. In 1947, the first human stereotactic apparatus was designed and used by Talairach in order to record and stimulate temporal structures in patients with epilepsy [4]. Cooper pioneered therapeutic chronic stimulation for epilepsy in the early 1970s with a focus on the cerebellum due to the existing evidence of the inhibitory effects of this structure.

An additional early target for epilepsy was the anterior nucleus of the thalamus (ANT). In fact, as early as 1979, Cooper implanted chronic ANT DBS electrodes in patients with drug-resistant focal impaired awareness seizures [5]. This very same structure has resurfaced recently in a double-blinded RCT showing clinical benefit in some adults with drug-resistant focal seizures and focal to bilateral tonic clonic seizures [6]. Velasco and colleagues in 1987 targeted the centromedian thalamic nuclei for treatment of generalized or multifocal uncontrollable seizures in five patients, where clinical seizures were significantly reduced as were EEG interictal spikes and slow waves [7]. The author performed further studies addressing the role of the centromedian thalamic nuclei in epilepsy pathogenesis and examining the long term effect of chronic electrical stimulation [8–10].

The technique of chronic stimulation of subcortical structures was proposed soon after the introduction of human stereotactic surgery in 1947 [11]. What was first used as a tool to study and treat neuropsychiatric disorders led to applications in pain management, then epilepsy, and finally movement disorders. The historical evidence presented challenges the notion of more recent published studies declaring DBS "a new approach" in epilepsy treatment [11].

2. Mechanism

Mechanisms of action of neuromodulation for epilepsy control are poorly understood and acknowledged as multifaceted and multifarious. Stimulation parameters used in clinical trials have been commonly experimental in nature, often derived from subjective configurations investigated with each anatomic structure in question. A chasm is still evident between insufficient animal data and limited clinical models. A need exists for more studies looking into the optimal stimulation parameters for the clinical management of seizures.

Early animal models by Ranck investigated the amounts of current necessary to stimulate various myelinated and unmyelinated neural structures within the central nervous system, showing that electrical fields have a variance in effect on different neuronal structures [12]. A proposed mechanism of action based on studies performed by Velasco and colleagues suggested that high frequency stimulation of kindled neuronal structures increases after-discharge thresholds with subsequent seizure reduction [13]. Additionally, it was noted that high frequency stimulation reduced regional cerebral blood supply in the stimulated area and in the centromedian thalamic nucleus, causing suppression of thalamic and cortical spike-wave and synchronous firing. Parahippocampal cortex high frequency stimulation was seen to enhance the GABAergic benzodiazepine receptor numbers in the operated field [13]. Interestingly, low-frequency stimulation had the potential to trigger or exacerbate epilepsy in some susceptible areas, but had an inhibitory effect in others [13,14].

Some authors speculate that direct electrical current may have inhibitory effects on neurons that participate in initiation, propagation, and protraction of epileptic activity in certain anatomical regions [15]. Such inhibitory effects may be due to high extracellular potassium accumulation post high frequency stimulation, causing depolarization of neurons and tonic inactivation of sodium channels, further prohibiting initiation of action potential and consequently seizure activation and propagation. Small elevations of potassium are capable of lowering the seizure threshold, but large increases cause an opposite reaction and thus inhibit pathological bursting [15].

Other important studies have shown that the mechanism of seizure inhibition may be more complex and possibly involves alteration of gene expression and protein synthesis [14]. The antiepileptogenic effects of low-frequency stimulation have been associated with the attenuation of adenosine receptor gene expression via inhibition of the dentate gyrus and chronic vagal nerve stimulation is seen to alter various amino acids and neurotransmitters in the brain, with decreases in aspartate (excitatory amino acid), increases in GABA (inhibitory), and increases in ethanolamine, a membrane lipid precursor [16].

Further mechanisms of action regarding each modulation intervention are described in their respective sections.

3. Modes of Treatment and Anatomical Targets for Stimulation in Epilepsy

Epilepsy affects 1% of the population worldwide, and approximately 30% of patients are drug-resistant [17,18]. Other than the few candidates for resective surgery, most will have persistent often disabling seizures for the rest of their lives [19]. Neuromodulation is an alternative treatment strategy for patient with drug-resistant epilepsy. It is most often used in those for whom resective surgery is not feasible (i.e., very extensive network, multifocal epilepsy, or epileptogenic zone in eloquent cortex).

There is much heterogeneity in the modes of electrical stimulation for the treatment of epilepsy, with a variety of anatomical structures, stimulation parameters, and outcome measures. Further discussion will cover structures and neuromodulation modes of treatment for epilepsy with a focus on published double-blinded RCTs (Table 1).

3.1. Deep Brain Stimulation (DBS)

Cerebellum: The earliest target of deep brain stimulation was the cerebellum. Therapeutic chronic stimulation became a treatment modality for epilepsy in the early 1970s pioneered by Cooper, and targeted due to the existing evidence of the inhibitory effects of this structure [20]. In his study, 10 of 15 patients with drug-resistant epilepsy had significant reduction or complete seizure inhibition during three years of chronic anterior lobe cerebellar stimulation [20]. According to Rosenow, in another group, Cooper targeted the anteromedial cerebellar surface for electrode placement and was able to achieve >50% reduction in seizure frequency in 18 of 32 patients [5]. Following these results, Van Buren presented a double-blinded cerebellar stimulation study of five patients with drug-resistant seizures that showed no significant difference in outcome [21]. Due to the resultant uncertainty surrounding the long-term outcomes of cerebellar stimulation, Wright et al. performed a double-blinded trial of chronic cerebellar stimulation in 12 patients with severe epilepsy [22]. Two 8-button pads were positioned on the upper surface of the cerebellum providing a mean peak current of 5–7 mA and frequency of 10 cathodal pulses per second of alternating polarity. Two of the patients had more bespoke parameters based on their responses. The patients received three modes of stimulation with randomly allocated two month phases of (1) continuous stimulation, (2) intermittent contingent stimulation, and (3) no stimulation. There was no significant reduction in seizure frequency in any of the groups within this trial [22]. After these results, cerebellar stimulation fell out of favor, until Velasco et al. re-evaluated the controversial topic with the aid of improved technology from radiofrequency-linked pulse generators to a fully implantable programmable battery-operated pulse generator [23]. Five drug-resistant epilepsy patients were involved in the study with insertion of two four-contact plates onto the supero-medial cerebellar surface. Fixed pulse width of 0.45 ms with current at 3.8 mA producing a charge density of $2.0 \ \mu C/cm^2$/phase was utilized. Pulse frequency was 10 pulses per second as used in the prior trials. The patients served as their own controls and in the initial three months double-blinded stage, a 33% reduction in seizures was reported in those with the stimulation initially on. All five patients in the unblinded stimulation period at six months had a mean seizure reduction rate of 41%, with significant reductions in tonic-clonic seizures and tonic seizures. Adverse events were all infectious in nature [23].

The differing target areas of stimulation and seizure patterns may be a reason for the variances seen in these studies. However, positive results seen in certain trials show that the cerebellum remains a potential target for neuromodulation in epilepsy.

Centromedian nucleus of the thalamus (CMT): As part of the cortico–striato–thalamic pathway, the CMT has extensive projections to the cortex and has been observed to be involved in cortical excitation and seizure propagation [24,25]. Several pioneering studies have shown seizure reduction and decreased frequency in EEG spiking for generalized epilepsy [7,26–29]; other case reports have also proven its clinical effectiveness in refractory status epilepticus [30,31].

Two double-blinded RCTs have been performed on this structure. Fisher et al. performed in six patients a cross-over on/off stimulation protocol in three month blocks with a three month washout period. A 30% reduction in tonic-clonic seizure frequency with stimulation vs. 8% reduction in the sham period was noted using stimulator parameters of 90 μs pulses at 65 pulses/s, 1 min of each 5 min, for 2 h/day. To maintain effective blinding, the stimulation amplitude was set to 50% of sensory threshold. No statistically significant improvement was seen in the double-blinded phase of study. However, thresholds were increased to 90% in the open phase follow-up of the investigation, where three of the six patients saw a >50% reduction in generalized seizure frequency. Velasco et al. performed in 13 patients a double-blinded 6 month cross-over protocol with 3 month period of no stimulation (between 6–12 months after implantation; when stimulating-alternating right and left paradigm using parameters of 60 Hz, 4–6 V. 1 min of each 5 min, for 24 h/day) [27]. No significance was seen in mean seizure frequency reduction. Long-term open-label follow-up however showed a mean seizure reduction of 81.6% in patients with Lennox-Gastaut syndrome compared to 57.3% in patients with focal epilepsy [27].

Anterior nuclei of the thalamus (ANT): As part of the limbic system, the ANT is connected to the hippocampus and receives projections from the mammillary bodies via the mamillothalamic tract and fornix, while itself projecting to the cingulate gyrus, orbito-frontal, and mesial prefrontal cortices [32]. Neuronal activity of the thalamic nuclei includes two major types of discharge: tonic and burst-firing. Additionally, theta activity is noted in some neurons and thought to play an important role in synaptic plasticity of the hippocampal circuit. Based on animal studies, and its central location with abundant connectivity, the ANT became a common and attractive target for DBS for the treatment of drug-resistant epilepsy. ANT DBS is an approved target for therapy in Europe in treating focal epilepsy for adults, whereas it is still awaiting approval from the United States Food and Drug Administration (FDA).

Cooper and Upton first published subjects who underwent ANT DBS for drug-resistant focal impaired awareness seizures in 1985. Five of six patients had a reduction of more than 60% in seizure frequency with stimulation at 3.5 V and 60–70 Hz [5]. Following this study, various case series reported a mean seizure reduction of >50% after ANT DBS [33–39]. These results subsequently led to a large double-blinded RCT for the Stimulation of the Anterior Nucleus of Thalamus for Epilepsy (SANTE) in patients with drug-resistant focal epilepsy, of which more than half of the study subjects had prior epilepsy or VNS surgery. Data from 110 patients with focal seizures focal to bilateral tonic-clonic seizures were collected [6]. Bilateral leads were implanted and after a one month post-operative baseline period, patients were randomized to receive three months of stimulation with duration 90 μs, frequency 145 Hz, at 5 V (on for one minute then off for five minutes) vs. no stimulation. After the double-blinded period, all participants received nine months of stimulation. Long-term follow-up showed a median seizure reduction from baseline at one year of 41% to 69% at five years. In the five years of follow-up, 16% of patients were seizure-free. Quality of life significantly improved from baseline at year one and five. Adverse effects included stimulation-related paresthesia (22.7%), implant site pain (20.9%), and infection (12.7%). Other open-label studies reported a reduction in seizure frequency >50%, however insertion effects could not be ruled out due to the nature of study design [36,38,40].

Hippocampus (HCP): Focal epilepsy involving the temporal lobe is well known to be the most resistant to pharmacological treatment. Even though surgical resection of the epileptic focus achieves more than 70% success rate in appropriately selected patients, this treatment option is not feasible for patients with bilateral disease or in those where resection would necessitate removal of the critical amygdalo-hippocampal complex responsible for verbal memory [18,41,42]. In such patients, the hippocampus serves as an appealing target for neuromodulation, and HCP sclerosis is known to be the most responsive to surgical treatment with favorable outcome [43,44].

Experiments on low-frequency amygdalo-hippocampal stimulation were initiated in the 1990s and showed profound effect on seizure development, expression, and thresholds, where it has been proposed that chronic low-frequency stimulation inhibits kindling [45,46]. Certainly, in long-term animal models, chronic low-frequency stimulation is associated with suppressed inhibitory effects over time and thus increased seizure thresholds [47,48]. Alternatively, a proposed mechanism for the effectiveness of high frequency hippocampal stimulation for epilepsy includes increased after-discharge thresholds and latencies that shorten the duration of the after discharges, which then reduce excitability as well as inducing hypoperfusion of the amygdalo-hippocampal complex [49–51].

Five RCTs evaluated hippocampal DBS [52–56]. The first by Tellez-Zenteno et al. performed a double-blind cross over RCT in four patients with refractory mesial temporal lobe epilepsy (MTLE) [52]. A median seizure reduction of 15% (not significant) was observed comparing stimulation vs. no stimulation. Seizure severity and symptomatology did not change [52]. Velasco et al. proceeded to a one month blinded trial in nine MTLE patients and reported a median seizure frequency reduction of 40% in the stimulation group vs. 0% in the no stimulation group (graphical presentation) [53]. However, neither numerical values were given nor were the statistical significance of the results stated [53]. McLachlan et al. also performed a cross-over RCT in two MTLE patients with mean reduction in seizures of 33% in those with stimulation [54]. Although, these three studies showed decreases in seizure frequency, sample sizes were too small and there was variability between the studies in placebo effect [52–54]. Thus, a larger trial was designed by Wiebe et al. who looked to examined whether hippocampal DBS was more safe and effective than simply implanting an electrode without stimulation [55]. Despite being a multicenter trial, only six out of a target sample of 57 were recruited leading to a halt in the RCT. None of the outcomes were statistically significant, likely due largely to the small sample size [55]. Cukiert et al. recently published a double-blinded RCT of hippocampal DBS in patients with drug-resistant temporal lobe epilepsy. Sixteen patients were randomized 1:1 to either stimulation or no stimulation. All patients received bipolar continuous stimulation with duration of 300 μs, frequency at 130 Hz and weekly 0.4 V stimulus intensity increments to a maximum of 2 V. A significant reduction in seizure frequency was observed in the stimulation group at full generator activation of 2 V. Half of the active group became seizure free. Seven of the eight participants had at least 50% reduction in seizure frequency. Local skin erosions were the main side-effects noted in this study [56]. Other open-label studies quoted seizure freedom rates of 15–45% [53,57–59].

The mechanisms involved in the reduction of seizures, using optimal stimulating parameters and targets are still unclear for hippocampal DBS. Further investigations into these variables are critical in delineating its potential benefit.

Nucleus Accumbens (NAc): This structure has an important role in both the anatomical and functional connectivity between frontal and temporal lobes [60]. In animal models, the NAc is seen to be involved in the propagation of epileptiform activity [61,62]. In the only RCT of this structure performed by Kowski et al. 4 patients underwent a cross-over protocol with bilateral DBS implantation of the NAc and ANT. One month post-surgery the patients were randomized to receive NAc stimulation or no stimulation (125 Hz, 5 V, 90 μs, 1 min stimulation/5 min off). The treatment protocol lasted three months and after a one month washout period the patients switched to the other protocol. The ANT was continuously switched on in all patients. Three out of the four patients experienced >50% reduction in frequency of disabling seizures with no further improvement with additional ANT stimulation [63]. These results will need to be further interrogated by higher powered studies in the future.

New DBS targets are continuously being identified and characterized for patients with difficult to treat epilepsy. The subthalamic nucleus (STN) has an important role in motor control and motor-related seizures and is thought to desynchronize motor pathways [64]. Several small case series exist of STN DBS inserted for cases of motor-related seizures with favorable outcome [65–67]. Following these small successes, a double-blinded RCT (STIMEP trial) was put in place. However, this was terminated due to insufficient enrollment. The caudate nucleus has also been seen as a target due to its involvement in the cortico–striato–thalamic pathway and rationale of inducing cortical hyperpolarization via neuromodulation [68,69]. Chkhenkeli et al. exhibited in a subset of their large cohort of patients that with low frequency 4–8 Hz stimulation, cortical and hippocampal interictal spiking and epileptiform activity decreased. However, due to the heterogeneity of the population, varying stimulation protocols, uncontrolled observations, and short follow-up it is difficult to properly interpret these studies.

3.2. Transcranial Direct Current Stimulation (tDCS)

tDCS is an emerging noninvasive stimulation technique that modulates cortical activity [70]. It utilizes weak direct current to modulate neuronal membrane potentials and hence cortical activity. The continuous stimulation in turn displaces polar-sensitive molecules, neurotransmitters, and receptors in brain tissue, triggering a polarity shift in membrane potentials [70,71]. Negative cathodal stimulation is proposed to cause cortical inhibition and diminish epileptiform discharges. The only double-blinded RCT compared cathodal tDCS at 2 mA for 30 min over the epileptic foci in three settings: three consecutive days, five consecutive days, and placebo stimulation [72]. In their 28 patients (3 day n = 12; 5 day n = 8; placebo n = 8), there was a significant reduction in seizure frequency at one and two months post-cathodal tDCS vs. baseline in all three arms of the study. There was significant mean seizure frequency reduction in both three and five day cathodal tDCS as compared with placebo at two months follow-up (48% reduction in the treatment group vs. 6.3% reduction in the placebo group). There was significantly increased reduction in seizures in both the three days cathodal tDCS group (43% reduction in the treatment group vs. 6.3% reduction in the placebo group and five days group (55% reduction in the treatment group vs. 6.3% reduction in the placebo group). Short-term interictal epileptiform discharges were also significantly reduced after stimulation in all groups. There were limited side effects with this mode of treatment, mostly local sensory discomfort and mild headaches [72].

tDCS can be used to manage both focal and generalized epilepsy in both children and adults and provides a slightly cheaper (vs. repetitive transcranial magnetic stimulation (rTMS)), portable and alternative mode of treatment especially in the younger population who cannot tolerate rTMS. The studies for tDCS are limited and the majority of investigations are preliminary, however much promise is seen.

3.3. Vagal Nerve Stimulation (VNS)

The first report of VNS was by Schwetzer and Wright in 1937, who looked at the effects of the knee jerk and various physiological changes in circulation and respiration by stimulating vagal afferents [73]. In 1997, the US FDA approved VNS as the first neuromodulation mode of treatment for drug-resistant epilepsy. In VNS surgery, the left vagus nerve, due to decreased cardiac side effects compared to the right, is stimulated. The stimulation of mostly (80%) afferent fibers is seen to converge onto the nucleus tractus solitarius, after which it proceeds onto the locus coeruleus [74]. Functional neuroimaging and electrophysiological studies of VNS have examined several areas of the brain (thalamus, cerebellum, orbitofrontal cortex, limbic system, hypothalamus, and medulla) and their levels of immediate response following stimulation. Albeit extensive, the clinical and animal VNS research remains inconclusive. It has been hypothesized that synaptic connections may be altered via VNS, modifying the electrical network in the brain [75]. Long-term VNS has been suggested to transform specific subcortical locations which in turn influence larger areas of the cortex [75–77].

Five RCTs investigating the efficacy of VNS were published between 1994 and 2014 [78–82]. The blinded phases of two key RCTs ultimately led to the approval of this mode of therapy by the FDA in epilepsy patients with drug-resistant epilepsy. These two RCTs showed that seizure frequency decreased >50% in 23–31% in the treatment groups vs. 13–15% in the control groups [78,79]. The study by Klinkenberg et al. is the only pediatric RCT to date [80]. This study demonstrated no significance in responder rate (>50% reduction in seizure frequency) with rates of 16% in high and 21% in low stimulation. One RCT compared VNS and best medical therapy vs. medical therapy alone, with no significant reduction in seizure frequency nor any difference in responder rates between treatment and control groups [81]. Aihua et al. explored the efficacy and safety of transcutaneous vagus nerve stimulation (tVNS) in patients with drug-resistant epilepsy. After 12 months, the monthly seizure frequency was lower in the tVNS group than in the control group (8.0 to 4.0; p = 0.003). All patients had improved Self-Rating Anxiety Scale, Self-Rating Depression Scale, Liverpool Seizure Severity Scale, and Quality of Life in Epilepsy Inventory-31 scores with minimal adverse effects including dizziness and drowsiness [82]. Other studies have stated complications of VNS placement included hoarseness (30%), dyspnea (13%), infection (12%), cough (7%), and throat pain (7%) [78–82].

3.4. Trigeminal Nerve Stimulation (TNS)

Following the treatment effects of VNS in epilepsy and evidence showing its mechanisms involving the locus ceruleus and nucleus solitarius, the trigeminal nerve was seen as a potential target since both the locus ceruleus and nucleus solitarius project onto the trigeminal nucleus [83–85]. External TNS (eTNS) is a noninvasive mode of treatment delivered by stimulating the trigeminal sensory roots within the facial tissue. Similarly to VNS, TNS is thought to produce an arousal-like effect via triggering the reticular activating system causing cortical and thalamic desynchronization [86].

DeGorgio et al. evaluated the safety and efficacy of eTNS in patients with drug-resistant epilepsy using a double-blind RCT design, to test the suitability of treatment and control parameters in preparation for a phase III multicenter clinical trial [87]. The responder rate, defined as >50% reduction in seizure frequency, was 30.2% for the eTNS group (120 Hz) vs. 21.1% for the active control group (2 Hz) for the 18-week treatment period (not significant, p = 0.31) [87]. The seizure frequency as measured by response ratio improved within each group compared to baseline. However, no differences were seen between the treatment and control groups. There was improvement in the patients' depression within and between groups (Beck Depression Inventory score change of −8.13 in treatment group vs. −3.95 in the control group). Although eTNS may not seem so efficacious, there are still advantages to this treatment that one should consider. It is non-invasive and more economical. This RCT provides preliminary evidence that eTNS is safe and may be effective in subjects with drug-resistant epilepsy. Side effects are primarily limited to anxiety, headache, and skin irritation. Patients may also benefit from dual stimulation due to the central connections of the crossed and uncrossed pathways of the trigeminal nerve and the links to key subcortical structures such as the locus ceruleus, ascending reticular activating system, and projections to other subcortical nuclei [88].

3.5. Repetitive Transcranial Magnetic Stimulation (rTMS)

Developed in 1985 in the United Kingdom, TMS was used to examine cortical excitability in various epilepsy syndromes, the antiepileptic medication effects on the brain and to interrogate areas of the brain for potential surgery [89]. Once rTMS could be used to excite or suppress neural activity for prolonged periods of time, studies began to look at its potential use of TMS as a treatment modality for epilepsy. TMS is an electrode free-electrical stimulator, which uses alternating magnetic fields in order to create electrical currents that in turn stimulate regional neurons to improve epilepsy symptomatology. The magnetic pulse stimulates a small area of cortical tissue, depolarizing nearby axons [90]. Repetitive stimulation is seen to lengthen the effects of depolarization, maintaining its effect for more than an hour post-treatment [91]. Other than targeting the specific area of interest via anatomical positioning over the patient's head, there is no way to target specific cell types, nor the interactions between inhibitory and excitatory cells. However, repetitive low-frequency stimulation has been hypothesized to cause prolonged inhibition as each pulse arrives during the late inhibitory phase of the last pulse, thus abating cortical hyper-excitability [90].

There has been a single double-blinded RCT of rTMS in patients with malformations of cortical development (MCD). Patients underwent five consecutive low-frequency (1 Hz, 1200 pulses) rTMS sessions targeting the MCD foci [92]. There was a significant reduction in seizure frequency in the rTMS group (−58% from baseline) vs. sham group (no change) as well as a significant reduction in epileptiform discharges straight after treatment (−31% from baseline) and at four weeks (−16% from baseline) vs. sham group (no change). There was improvement in subjective measures of social interaction and energy level and cognition in the treatment group vs. control. No serious adverse effects were reported. Headaches were experienced both in the treatment (25%) and control (22%) groups. There was no worsening of seizures. One patient in the control group reported insomnia [92]. Other open-label studies have shown reductions in seizure frequency with pathologies near the cortex in frontal and centro-parietal epilepsy, whereas targeting deeper structures in mesial temporal epilepsy has been associated with poor efficacy [93–95]. The majority of studies however have only shown evidence of decreasing epileptiform discharges, with no significant change in seizure reduction. A reason for such inconsistencies can be due to the selection bias, blinding bias, and differences in stimulation parameters [93–95].

rTMS is a non-invasive, inexpensive, pain-free procedure that has can modulate cortical brain activity. Evidence shows that rTMS is effective at abating epileptiform discharges, however the evidence for seizure reduction is still inconclusive. rTMS can certainly provide an alternative mode of treatment when considering foci lying over eloquent cortex that is not appropriate for surgery. However, further well-designed RCTs looking into the efficacy of this treatment are required to determine optimal stimulation frequencies, duration of treatment, intensity, and even the shapes of the magnet we use.

3.6. Responsive Neurostimulation (RNS)

The FDA approved RNS System (NeuroPace, Mountain View, CA, USA) is the first intracranial closed-loop system providing responsive stimulation directly to one or two seizure foci. Real time abnormal electrographic activity is detected and an automatic responsive stimulation is triggered thus halting any evolving seizure activity from propagating. Detection and stimulation parameters may be adjusted according to clinical benefit and to minimize side-effects related to stimulation. With the aforementioned possible mechanisms of stimulation in treating epilepsy, the RNS system is more amenable to the physiological transformations occurring during stimulation, thus potentially providing a better treatment modality over nonresponsive or continuous stimulation.

In a large multi-centered double-blinded RCT, 191 patients who had drug-resistant focal seizures (defined in this RCT as failed $\geq$2 antiepileptic medication trials, $\geq$3 seizures/month, and 1 or 2 seizure foci) were implanted with an RNS system [96]. Morrell et al. reported 37.9% reduction in seizure frequency in the treatment arm vs. 17.3% reduction in the sham group at the end of the blinded phase [76]. However, no difference in responder rates was seen between the treatment and sham groups during the blinded phase of the study; additional reductions in seizure frequency in the treatment group to 44% at one year and 53% at two years were reported during the open-label extension [96,97]. Complications of RNS include a 4.7% rate of intracerebral hemorrhage and a 9% rate of infection after a mean of 5.4 years of follow-up, requiring neurostimulator explantation in 4.7% of the cases [96,97]. The six-year long-term analysis from this trial has shown that RNS mitigated substantial and sustained seizure reduction in their cohort of 111 patients with drug-resistant mesial temporal lobe epilepsy [98]. Using last observation carried forward (LOCF) analyses a median of 70% (interquartile range 31.8–92.9%; n = 106) seizure reduction was seen, and 50% responder rates of 66% (95% CI 56.6–74.4%) at six years. Forty-five percent (50/111) of patients reporting a seizure free period of $\geq$3 months, 29% (32/111) of patients $\geq$6 months, and 15% (17/111) of patients $\geq$ 1 year. It was also noted that the seizure reduction did not correlate with clinical characteristics such as mesial temporal sclerosis, bilateral seizure onset, and prior respective surgery/VNS operation. This treatment continues to improve with each year of implantation, and thus there is true potential for true neuromodulation and slow improvement [98]. In this same cohort, seizure reduction response in those with partial-onset seizures arising from eloquent cortex was investigated [99]. Over two-to-six year period post-implantation, a median seizure reduction of 70% was seen in frontal onset seizures, 58% in temporal neocortex and 51% in those of multilobar onset (LOCF analysis). It was also noted that therapeutic stimulation of eloquent cortex could be given sub-threshold and not exhibit side effects such as involuntary motor movement of altered motor performance when stimulating the primary motor cortex [99].

Table 1. Summary of double-blinded randomized controlled trials of neuromodulation in the management of drug resistant epilepsy.

Study, Year	Country	Intervention	Study Setting	Population	Results	Follow Up	Complications
				Randomized Controlled Trials of VNS			
Ben-Menachem et al. 1994 [78]	Sweden	High vs. low stimulation treatment	Multicenter, Children and adults	- Drug resistant seizures - Numbers—High: n = 54 vs. Low: n = 60 - Mean age—High: 33.1 years vs. Low: 33.5 years - Females—High: 39% vs. Low: 37% - Mean duration—High: 23.1 years vs. Low: 20 years	- Reduction in seizure frequency—High: 24.5% vs. Low: 6.1% (p = 0.01) - At least 50% reduction in seizure frequency—High: 31% vs. Low: 13% (p = 0.02) - No seizure free patients - Seizure types were not significant - Implant was well tolerated	- 12 weeks post 2-week recovery period from surgical implantation	- Hoarseness (33%) - 1 patient died from myocardial infarction - 1 patient developed total vocal cord paralysis
Handforth et al. 1998 [79]	USA	High vs. low stimulation treatment	Multicenter, Children and adults	- Drug resistant focal impaired awareness seizures (patient had at least focal seizures over 30 days or focal seizures to bilateral tonic-clonic seizures) - Numbers—High: n = 95 vs. Low: n = 103 - Mean age—High: 32.1 years vs. Low: 34.2 years - Females—High: 48% vs. Low: 57% - Mean duration—High: 22.1 years vs. Low: 23.7 years	- Reduction in seizure frequency—High: 27.9% vs. Low: 15.2% (p = 0.04) - No difference in between-group comparison for 50% responders—High: 23.4% vs. Low: 15.7% - One patient seizure free (High) - No change in physiologic indicators of cardiac or pulmonary function	- 12–16 weeks after 2-week ramp-up period	- Hoarseness (30%) - Dyspnea (13%) - Infection (12%)
Klinkenberg et al. 2012 [80]	Netherlands	High vs. low stimulation treatment for 20 weeks, then all received high for 19 weeks	Single center, Children	- 41 children total (35 with focal epilepsy: 25 structural, 10 unknown etiology; 6 with generalized epilepsy) - Numbers—High: n = 21 vs. Low: n = 20 - Mean age—High: 10 years 11 months vs. Low: 11 years 6 months - Mean duration—High: 7 years 8 months vs. Low: 9 years 5 months - Seizure frequency and severity were recorded using diaries and the adapted Chalfont Seizure Severity Scale	- Reduction in seizure frequency at least 50%—High: 16% vs. Low: 21%	- 20 weeks	- Voice alteration (20%) - Coughing (7%) - Throat pain (7%) - Infection (5%)

Table 1. *Cont.*

Randomized Controlled Trials of VNS

Study, Year	Country	Intervention	Study Setting	Population	Results	Follow Up	Complications
Rylin et al. 2014 [81]	France	PuLsE (Open Prospective Randomized Long-Term Effectiveness): VNS + Best medical practice (BMP) vs. BMP	Multicenter, Adults—early termination of trial due to low enrollment	- Drug resistant focal seizures (available baseline data and ≥1 post-op QOLIE-89: 48 with VNS + BMP, and 48 with BMP alone) - Mean age—Treatment group: 38 years vs. Control group: 41 years - Females—Treatment group: 50% vs. Control group: 44% - Mean duration—Treatment group: 25 years vs. Control group: 25 years	- Significant improvements in HRQoL (QOLIE-89)—Treatment group (VNS+BMP): 5.5 points vs. Control group (BMP): 1.2 ($p < 0.05$) - No difference in secondary endpoints: 1. Seizure frequency 2. Responder rate 3. CES-D 4. NDDI-e 5. AEP - AED Load	- 24 months in seven patients, 12 months in 60.	- Transient vocal cord paralysis (4%) - Brief period of respiratory arrest (3%)
Aihua et al. 2014 [82]	China	Transcutaneous: Ramsay Hunt zone stimulation (treatment group) vs. earlobe (control stimulation)	Single center, Children and adults	- Numbers—60 patients randomly divided into two groups based on stimulation zone - Mean age—Treatment group: 34.5 years vs. Control group: 29.0 years - Mean duration—Treatment group: 10.7 years vs. Control group: 17.6 years - Seizures types—Treatment group: focal aware (65%), focal impaired awareness (23%), generalized (11%) vs. Control group: focal onset aware (71%), focal impaired awareness (14%), generalised (14%)	- Reduction in seizure frequency at 12 months—Treatment group: 8/month vs. Control group: 4/month ($p = 0.003$) - Antiepileptic drugs were maintained at a constant level in all subjects. - All patients showed improved SAS, SDS, LSSS, QOLIE-31 scores	- 12 months	- Dizziness (3%) - Drowsiness (9%)
Randomized controlled trials of DBS							
Van Buren et al. 1978 [21]	USA	Bilateral stimulation of the superior surface of the cerebellum. Treatment group: 10–14 V, 10 Hz vs. off stimulation	Single center, Adults	- Drug-resistant epilepsy - Numbers—5 - Mean age—27.2 years (18–34) - Mean duration—8 to 23 years	- No significant differences in seizure frequency were identified	- Up to 1 or more weeks (total 52 days) of blinded phase over 15–21 months	- No complications mentioned

Table 1. *Cont.*

		Randomized Controlled Trials of VNS				
Study, Year	Country	Intervention	Study Setting	Population	Results	Follow Up	Complications
Wright et al. 1984 [22]	United Kingdom	Stimulation of the upper surface of the cerebellum 2 cm from midline on each side Treatment group: 1–7 mA, 10 Hz in either continuous or contingent session vs. sham stimulation	Single center, Adults	- Drug-resistant epilepsy - Numbers—12 - Mean age—30 years (20–38) - Female—17% - Mean duration—10 to 32 years	- No reduction in seizure frequency occurred that could be attributed to stimulation - Cerebellar stimulation is not recommended	- 6-month blinded phase consisted of 3 × 2 month periods	- Infection with electrodes removal (16.7) - Electrode displacement required reoperation (25%) - Lead pain that required repositioning (8.3%) - Apparatus failure (8.3%) - Receiver pocket burst (8.3%)
Velasco et al. 2005 [23]	Mexico	Cerebellar stimulation -bilateral modified four-contact plate electrodes adjusted to 2.0 μC/cm² / phase	Single center, Adults	- Drug-resistant focal motor seizures - Numbers—*n* = 5 (*n* = 3 with stimulation ON and *n* = 2 with stimulation OFF in blinded phase) - Randomized blinded phases for 3 months followed by all ON stimulation - Patients served as own controls (Compared seizure frequency pre-implant (3 months) vs. post-implant phases (average, eight epochs of 3 months each)	- Reduction in seizure frequency—Stimulation ON: GTCs to 33% vs. OFF: no change (at 3 months) (patient 2, 21%; patient 3, 46%; patient 4, 32%) (*p* = 0.023) Open label for 6 months: - Mean seizure rate of 41% of the baseline	- 3 months	- 1 infection that required implant removal
Fisher et al. 1992 [26]	USA	Bilateral stimulation of the centromedian thalamic nucleus (0.5 to 10 V, 65 Hz, 90 μs pulse width) vs. sham stimulation	Multicenter, Adults	- Drug-resistant focal epilepsy (1), focal epilepsy with generalization (1), generalized epilepsy (5) - Numbers—7 - Mean age—28 - Female—57.1% - Mean duration—14 to 29 years	- Reduction in seizure frequency—treatment group: 30% reduction vs. 8% sham stimulation - During open label time, 3 patients reported 50% decrease in seizure frequency	- 3 months period of either treatment or sham stimulation with a 3-month washout phase between Followed by 3–13 months open label with all on	- No complications mentioned

Table 1. *Cont.*

Study, Year	Country	Intervention	Study Setting	Population	Results	Follow Up	Complications
					Randomized Controlled Trials of VNS		
Velasco et al. 2000 [27]	Mexico	Alternate stimulation between the left and the right centromedian thalamic nucleus (4–6 V, 60 Hz, 450 µs pulse width) vs. sham stimulation	Single center, Adults and children	- Lennox-Gastaut syndrome with atypical absences and GTCS (8), complex partial and secondary generalized (5) - Numbers—13 - Mean age—19.2 years - Female—38% - Mean duration—4 to 33 years	- Total number of seizure, GTCS, absence and CPS were reduced significantly (absolute and relative percentage decrease) turning off stimulation did not cause a return to base line levels - No statistical significant differences between the on and off periods ($p = 0.23$) - 5 out of 11 patients with GTCS and 2 out of 8 patients with absence seizures were seizure free	- 6–9 months period of stimulation followed by a 6-month cross-over pairs: 2 × 3 month phase On./Off or Off/On - Followed by 42 months follow up	- 1 death due to herpes encephalitis
Fisher et al. 2010 [6]	USA	Anterior nuclei of thalamus stimulation	Multicenter, Adults	- Drug resistant focal seizures - Numbers—Treatment group = 54 vs. Control group = 5 - Mean age—Treatment group: 35.2 years vs. Control group: 36.8 years - Female—Treatment group: 54% vs. Control group: 46% - Mean duration—Treatment group: 21.6 years vs. Control group: 22.9 years	- Reduction in seizure frequency—Treatment group: 29% greater reduction vs. control group (last month of blinded trial) - Responder rate 54% by 2 years - 14 patients seizure free for at least 6 months	- 3 months blinded followed by 9 months open label with all on - 5-year follow-up study (Salanova)	- Paresthesias (22.7%) - Implant site pain (20.9%) - Implant site infection (12.7%)
Tellez-Zenteno et al. 2006 [52]	Canada	Left hippocampal stimulation (1.8 V to 4.5 V, 190 Hz, 90 µs pulse width) vs. sham stimulation	Single center, Adults	- Drug-resistant left unilateral MTLE - Numbers—4 - Mean age—31.8 years - Female—75% - Mean duration—16 to 24 years	- A median reduction in seizure frequency during treatment—15% - Seizure improved in three patients, but the result was not statistically significant	- 3 × 2-month treatment pairs with monthly phase On or Off	- No complications mentioned

Brain Sci. **2018**, *8*, 69

Table 1. *Cont.*

				Randomized Controlled Trials of VNS			
Study, Year	Country	Intervention	Study Setting	Population	Results	Follow Up	Complications
Velasco 2007 [53]	Mexico	Bilateral or unilateral hippocampal stimulation Treatment group: 130 Hz, 450 μs pulse width vs. control group: No stimulation	Single center, Children and adults	- Drug-resistant CPS and GTCS - Numbers—Treatment group: n = 4 vs. Control group: n = 5 - Age—Treatment group: 20–40 years vs. Control group: 14–43 years - Female—Treatment group: 25% vs. Control group: 40% - Mean duration—Treatment group: 12.0 years vs. Control group: 10.4 years	- Seizure reduction in treatment group vs. baseline seizure frequency in control group, proving that the initial seizure decrease is not due to electrode implantation effect. - 18 months follow up: >95% seizure reduction in 5 patients with normal MRI, and 50–70% seizure reduction in 4 patients with hippocampal sclerosis	- 1-month blinded phase followed by 18–84 months open label with all on	Skin erosion with local infection in 3 patients, one of which required plastic surgery and eventual electrode removal
McLachlan et al. 2010 [54]	Canada	Bilateral hippocampal stimulation Treatment group: 185 Hz, 90 μs pulse width	Single center, Adults	- Drug-resistant focal epilepsy with bitemporal origination - Numbers—2 - Age—45 and 54 years - Female—50% - Mean duration—15 and 29 years	- Reduction in seizure frequency by 33% in the two patients during stimulation - Reduction in seizure frequency by 25% for the 3 months post stimulation before return to baseline (p < 0.01)	- 3 months period of stimulation On/Off followed by 3 months washout and repeat cycle	No complications noted
Wiebe et al. 2013 [55]	Canada	Hippocampal stimulation, unilateral or bilateral (Treatment group: 135 Hz continuous cathodal stimulation of all electrodes involved in seizure generation vs. control group: no stimulation)	Multicenter, Adult	- Drug-resistant MTLE - Numbers—Treatment group: 2 patients vs. Control group: 4 patients - Mean age—Treatment group: 30 years vs. Control group: 35–46 - Baseline seizure frequency—Treatment group: 12 seizures per month vs. Control group: 10 seizures per month	- Statistically nothing significant - Mean seizure reduction: Treatment group: 45% vs. Control group: 60% increase - Half of the patients in treatment group had >50% reduction. - Improvement with hippocampus sclerosis in the frequency of all types of seizures, and in subjective memory function - Borderline significant improvement in attention/concentration - Recall function worse	- 7 months	No complications mentioned

Table 1. *Cont.*

Study, Year	Country	Intervention	Study Setting	Population	Results	Follow Up	Complications
				Randomized Controlled Trials of VNS			
Cukiert et al. 2017 [56]	Brazil	Hippocampal stimulation, unilateral or bilateral (Treatment group: active stimulation at continuous 130 Hz, duration 300 μs, final intensity of 2 V (0.4 V increments) vs. control group: no stimulation)	Single center, Children and adults	- Drug resistant temporal lobe epilepsy - Numbers—Treatment group: 8 patients vs. Control group: 8 patients - Mean age—38.4 years - Mean pre-operative seizure frequency 12.5/month	- Treatment group: 50% of patients seizure free, 87.5% had >50% seizure reduction - Significant reduction in seizures (focal impaired awareness) from first month to end of blinded phase - Significant reduction in seizures (focal awere) except 3rd month of blinded phase	6 months blinded phase	Local skin erosion (12.5%)
Kowski et al. 2015 [63]	Germany	Bilateral stimulation of nucleus accumbens and the anterior thalamic nuclei Treatment group: 5 V, 125 Hz, 90 μs pulse width	Single center, Adults	- Inclusion criteria: 3 major seizures every 4 weeks during 3 month period - Numbers—4 - Mean age—36.7 - Female—75% - Mean duration—12.5 years	- Reduction in seizure frequency >50% in 3 patients	- 3 months period of stimulation On/Off, followed by 1 month washout and repeat cycle - 3-month open label - Anterior thalamic nuclei always stimulated	1 infection that required implant removal, but a patient re-participated in the study after clearance of infection
San-Juan et al. 2017 [72]	Mexico	Transcranial direct current stimulation (tDCS)–(randomized into three treatment arms: 2 mA cathodal direct current stimulation for 30 min: (1) three days (2) five days days vs. (3) placebo)	Multicenter, Adults	- Drug-resistant MTLE with hippocampal sclerosis - Numbers—n = 28 - Mean age—37.8 years	- Reduction in seizure frequency—Treatment groups: 48% vs. Placebo group: 6.3% (at 2 months) (p = 0.008)	2 months	2 patients had a focal impaired awareness seizure towards the end of first day session, however not believed to be intervention-related given high baseline seizure frequency

Table 1. *Cont.*

Study, Year	Country	Intervention	Study Setting	Population	Results	Follow Up	Complications
				Randomized Controlled Trials of VNS			
Velasco et al. 2005 [23]	Mexico	Cerebellar stimulation-bilateral modified four-contact plate electrodes adjusted to 2.0 μC/cm^2/phase	Single center, Adults	- Drug-resistant focal motor seizures - Numbers—$n = 5$ ($n = 3$ with stimulation ON and $n = 2$ with stimulation OFF in blinded phase) - Randomized blinded phases for 3 months followed by all ON stimulation - Patients served as own controls (Compared seizure frequency pre-implant (3 months) vs. post-implant phases (average, 8 epochs of 3 months each)	- Reduction in seizure frequency—Stimulation ON: GTCs to 33% vs. OFF: no change (at 3 months) (patient 2, 21%; patient 3, 46%; patient 4, 32%) ($p = 0.023$) Open label for 6 months: - Mean seizure rate of 41% of the baseline	3 months	- 1 infection that required implant removal
				Randomized controlled trial of TNS			
DeGiorgio et al. 2013 [87]	USA	External trigeminal nerve stimulation (eTNS)-treatment group: eTNS 120 Hz vs. control group: eTNS 2 Hz	Multicenter, Adults	- At least 2 focal seizures/month - Numbers—Treatment group: $n = 25$ vs. Control group: $n = 25$ - Mean age—Treatment group: 33.1 years vs. Control group: 34 years - Female—Treatment group: 64% vs. Control group: 44% - Mean duration—Treatment group: 16.7 years vs. Control group: 12.0 years	- No difference in responder rate—Treatment group: 31% vs. Control group: 21.1% ($p > 0.05$) - Improved seizure frequency within each group as measured by response ratio, but no difference between treatment vs. control group - Improvement in depression—Treatment group: BDI score change of −8.13 vs. Control group: BDI score change of −3.95 ($p = 0.002$)	18 weeks	- Skin irritation (14%) - Anxiety (4%) - Headache (4%)
				Randomized controlled trial of rTMS			
Fregni et al. 2006 [92]	USA	Repetitive transcranial magnetic stimulation (rTMS)-treatment group: 1 Hz, 1200 pulses vs. sham group	Single center, Adults	- MCD - Numbers—Treatment group: $n = 12$ vs. Sham group: $n = 9$ - Mean age—Treatment group: 21.3 vs. Sham group: 22.7	- Reduction in seizure frequency—Treatment group: 58% reduction vs. Sham group: No difference from baseline - In treatment group only: significant decrease in the number of epileptiform discharges immediately post treatment ($p = 0.01$) and at week 4 ($p = 0.03$)	60 days	- Headache (Treatment group: 5% vs. Sham group: 22%) - Insomnia (11%)

Table 1. *Cont.*

Study, Year	Country	Intervention	Study Setting	Population	Results	Follow Up	Complications
				Randomized Controlled Trials of VNS			
				Randomized controlled trial of RNS			
Morrell et al. 2011 [96]	USA	Responsive neuromodulation	Multicenter, Adults	- Failed ≥2 antiepileptic medication trials, ≥3 seizures/month, and 1 or 2 seizure foci - Numbers—Active stimulation: n = 97 vs. Sham stimulation n = 94 - Mean age—Active stimulation: 34.0 vs. Sham stimulation: 35.9 - Female—Active stimulation 48% vs. Sham stimulation 47% - Mean duration—Active stimulation: 20.0 years vs. Sham stimulation 21.0 years	- Reduction in seizure frequency—Active stimulation: −37.9% vs. Sham stimulation: −17.3% (*p* = 0.012) during blinded period - Responder rate—Active stimulation: 29% vs. Sham stimulation: 27% - 2 cases in active stimulation group were seizure free for the blinded phase - QOLIE-89 scores improved in active and sham stimulation, continued through 1 and 2 years **Open label:** - Median % reduction in seizure frequency of 44% (1st year), 53% (2nd year) - Statistically significant improvement in QOLIE scales at 1 and 2 years (*p* < 0.05)	12-week blinded period followed by 84-week open-label period	- Intracerebral hemorrhage (4.7%) - Infection (5.2% at end of open-label phase, 9.0% after mean follow-up 5.4 years of which 4.7% underwent explantation)

VNS (Vagus Nerve Stimulation); DBS (Deep Brain Stimulation); TNS (Trigeminal Nerve Stimulation); rTMS (Repetitive Transcranial Magnetic Stimulation); RNS (Responsive Neurostimulation); CES-D (Center for Epidemiologic Studies Depression Scale); NDDI-e (Neurological Disorders Depression Inventory–Epilepsy); AEP (Adverse Event Profile); AED (Antiepileptic Drug); SAS (Self-Rating Anxiety Scale); SDS (Self-Rating Depression Scale); LSSS (Liverpool Seizure Severity Scale); MTLE (Mesial Temporal Lobe Epilepsy); MCD (Malformation of Cortical Development); QOLIE-31 (Quality of Life in Epilepsy Inventory); BMP (Best Medical Practice); HRQoL (Health-Related Quality Of Life).

In addition, some recent open label studies with chronic continuous stimulation have also exhibited good outcomes [100–102]. Child et al. produced a proof of concept paper to show that continuous neocortical neurostimulation could provide significant reduction in seizure frequency, especially in those ineligible for resective surgery due to focal epilepsy arising from eloquent cortex [100]. Valentin et al. exhibited >90% seizure frequency reduction with one patient experiencing resolution of epilepsia partialis continua [101]. Lundstrom et al., in their cohort of 13 patients that underwent subthreshold cortical stimulation, showed suppression of interictal epileptiform discharges and improvement in clinical seizures [102].

RNS is a promising treatment for focal epilepsies even in bilateral disease and seizure foci affecting eloquent cortex. RNS, in comparison to the open-loop DBS and VNS, has a longer battery life due to a lower dose of stimulation. Stimulation is also not felt by patients, even when placed in an eloquent area, due to its low intensity. RNS is generally offered to those patients who are not suitable for resective surgery and who have foci in one or two areas of the brain. Further studies are necessary to optimize seizure detection, improve stimulation parameters, and build more contacts for cases of multiple foci.

4. Conclusions

Neuromodulation is a treatment strategy that is being used increasingly in those suffering from drug-resistant epilepsy that is not suitable for resective surgery. We are seeing more double-blinded RCTs demonstrating the efficacy of neurostimulation seizure patients. Although reductions in epilepsy frequency and focus firing are common in these trials, obtaining seizure freedom is rare. Invasive neuromodulation procedures (DBS, VNS, and RNS) have been approved as treatment measures. However, further investigations are necessary to delineate effective targeting, minimize side effects that are related to chronic implantation and to improve the cost effectiveness of these devices. The RCTs involved in the non-invasive modes of treatment whilst showing much promise (tDCS, eTNS, rTMS), have had only small recruitment numbers within their trials. Thus, they have not been sufficiently large enough to provide strong evidence on efficacy. Certainly, larger studies are needed, as well as studies that focus on better targeting techniques.

Acknowledgments: The authors have not received funding for this research and its publication.

Author Contributions: C.S.K. and N.J. conceived and designed the project. All authors contributed to the writing of this manuscript.

Conflicts of Interest: The authors declare no conflict of interest.

References

1. Spiegel, E.A.; Wycis, H.T.; Marks, M.; Lee, A.J. Stereotaxic Apparatus for Operations on the Human Brain. *Science* **1947**, *106*, 349–350. [CrossRef] [PubMed]
2. Gildenberg, P.L. Evolution of neuromodulation. *Stereotact. Funct. Neurosurg.* **2005**, *83*, 71–79. [CrossRef] [PubMed]
3. Sheer, D.E. *Electrical Stimulation of the Brain: An Interdisciplinary Survey of Neurobehavioral Integrative Systems*; University of Texas Press: Austin, TX, USA, 1961.
4. Talairach, J.; Hecaen, H.; David, M.; Monnier, M.; Deajuriaguerra, J. Recherches Sur La Coagulation Therapeutique Des Structures Sous-Corticales Chez Lhomme. *Rev. Neurol.* **1949**, *81*, 4–24.
5. Rosenow, J.; Das, K.; Rovit, R.L.; Couldwell, W.T.; Irving, S. Cooper and his role in intracranial stimulation for movement disorders and epilepsy. *Stereotact. Funct. Neurosurg.* **2002**, *78*, 95–112. [CrossRef] [PubMed]
6. Fisher, R.; Salanova, V.; Witt, T.; Worth, R.; Henry, T.; Gross, R.; Oommen, K.; Osorio, I.; Nazzaro, J.; Labar, D.; et al. Electrical stimulation of the anterior nucleus of thalamus for treatment of refractory epilepsy. *Epilepsia* **2010**, *51*, 899–908. [CrossRef] [PubMed]
7. Velasco, F.; Velasco, M.; Ogarrio, C.; Fanghanel, G. Electrical stimulation of the centromedian thalamic nucleus in the treatment of convulsive seizures: A preliminary report. *Epilepsia* **1987**, *28*, 421–430. [CrossRef] [PubMed]

8. Velasco, F.; Velasco, M.; Marquez, I.; Velasco, G. Role of the centromedian thalamic nucleus in the genesis, propagation and arrest of epileptic activity. An electrophysiological study in man. *Acta Neurochir. Suppl.* **1993**, *58*, 201–204. [PubMed]

9. Velasco, F.; Velasco, M.; Velasco, A.L.; Jimenez, F. Effect of chronic electrical stimulation of the centromedian thalamic nuclei on various intractable seizure patterns: I. Clinical seizures and paroxysmal EEG activity. *Epilepsia* **1993**, *34*, 1052–1064. [CrossRef] [PubMed]

10. Velasco, F.; Velasco, M.; Velasco, A.L.; Jimenez, F.; Marquez, I.; Rise, M. Electrical stimulation of the centromedian thalamic nucleus in control of seizures: Long-term studies. *Epilepsia* **1995**, *36*, 63–71. [CrossRef] [PubMed]

11. Schulze-Bonhage, A. Deep Brain Stimulation: A New Approach to the Treatment of Epilepsy. *Dtsch. Arztebl. Int.* **2009**, *106*, 407–412. [PubMed]

12. Ranck, J.B., Jr. Which elements are excited in electrical stimulation of mammalian central nervous system: A review. *Brain Res.* **1975**, *98*, 417–440. [CrossRef]

13. Velasco, F.; Velasco, A.L.; Velasco, M.; Carrillo-Ruiz, J.D.; Castro, G.; Trejo, D.; Nunez, J.M. Central nervous system neuromodulation for the treatment of epilepsy I-Efficiency and safety of the method. *Neurochirurgie* **2008**, *54*, 418–427. [CrossRef] [PubMed]

14. Jahanshahi, A.; Mirnajafi-Zadeh, J.; Javan, M.; Mohammad-Zadeh, M.; Rohani, R. The antiepileptogenic effect of electrical stimulation at different low frequencies is accompanied with change in adenosine receptors gene expression in rats. *Epilepsia* **2009**, *50*, 1768–1779. [CrossRef] [PubMed]

15. Boon, P.; Raedt, R.; de Herdt, V.; Wyckhuys, T.; Vonck, K. Electrical stimulation for the treatment of epilepsy. *Neurotherapeutics* **2009**, *6*, 218–227. [CrossRef] [PubMed]

16. Ben-Menachem, E.; Hamberger, A.; Hedner, T.; Hammond, E.J.; Uthman, B.M.; Slater, J.; Treig, T.; Stefan, H.; Ramsay, R.E.; Wernicke, J.F.; et al. Effects of vagus nerve stimulation on amino acids and other metabolites in the CSF of patients with partial seizures. *Epilepsy Res.* **1995**, *20*, 221–227. [CrossRef]

17. Fiest, K.M.; Sauro, K.M.; Wiebe, S.; Patten, S.B.; Kwon, C.S.; Dykeman, J.; Pringsheim, T.; Lorenzetti, D.L.; Jette, N. Prevalence and incidence of epilepsy: A systematic review and meta-analysis of international studies. *Neurology* **2017**, *88*, 296–303. [CrossRef] [PubMed]

18. Kwan, P.; Brodie, M.J. Early identification of refractory epilepsy. *N. Engl. J. Med.* **2000**, *342*, 314–319. [CrossRef] [PubMed]

19. Engel, J., Jr.; Wiebe, S.; French, J.; Sperling, M.; Williamson, P.; Spencer, D.; Gumnit, R.; Zahn, C.; Westbrook, E.; Enos, B.; et al. Practice parameter: Temporal lobe and localized neocortical resections for epilepsy: Report of the Quality Standards Subcommittee of the American Academy of Neurology, in association with the American Epilepsy Society and the American Association of Neurological Surgeons. *Neurology* **2003**, *60*, 538–547. [PubMed]

20. Cooper, I.S.; Amin, I.; Riklan, M.; Waltz, J.M.; Poon, T.P. Chronic cerebellar stimulation in epilepsy. Clinical and anatomical studies. *Arch. Neurol.* **1976**, *33*, 559–570. [CrossRef] [PubMed]

21. Van Buren, J.M.; Wood, J.H.; Oakley, J.; Hambrecht, F. Preliminary evaluation of cerebellar stimulation by double-blind stimulation and biological criteria in the treatment of epilepsy. *J. Neurosurg.* **1978**, *48*, 407–416. [CrossRef] [PubMed]

22. Wright, G.D.; McLellan, D.L.; Brice, J.G. A double-blind trial of chronic cerebellar stimulation in twelve patients with severe epilepsy. *J. Neurol. Neurosurg. Psychiatry* **1984**, *47*, 769–774. [CrossRef] [PubMed]

23. Velasco, F.; Carrillo-Ruiz, J.D.; Brito, F.; Velasco, M.; Velasco, A.L.; Marquez, I.; Davis, R. Double-blind, randomized controlled pilot study of bilateral cerebellar stimulation for treatment of intractable motor seizures. *Epilepsia* **2005**, *46*, 1071–1081. [CrossRef] [PubMed]

24. Mondragon, S.; Lamarche, M. Suppression of Motor Seizures after Specific Thalamotomy in Chronic Epileptic Monkeys. *Epilepsy Res.* **1990**, *5*, 137–145. [CrossRef]

25. Miller, R. Cortico-thalamic interplay and the security of operation of neural assemblies and temporal chains in the cerebral cortex. *Biol. Cybern.* **1996**, *75*, 263–275. [CrossRef] [PubMed]

26. Velasco, A.L.; Velasco, F.; Jimenez, F.; Velasco, M.; Castro, G.; Carrillo-Ruiz, J.D.; Fanghanel, G.; Boleaga, B. Neuromodulation of the centromedian thalamic nuclei in the treatment of generalized seizures and the improvement of the quality of life in patients with Lennox-Gastaut syndrome. *Epilepsia* **2006**, *47*, 1203–1212. [CrossRef] [PubMed]

27. Velasco, F.; Velasco, M.; Jimenez, F.; Velasco, A.L.; Brito, F.; Rise, M.; Carrillo-Ruiz, J.D. Predictors in the treatment of difficult-to-control seizures by electrical stimulation of the centromedian thalamic nucleus. *Neurosurgery* **2000**, *47*, 295–304; discussion 304–305. [CrossRef] [PubMed]

28. Cukiert, A.; Burattini, J.A.; Cukiert, C.M.; Argentoni-Baldochi, M.; Baise-Zung, C.; Forster, C.R.; Mello, V.A. Centro-median stimulation yields additional seizure frequency and attention improvement in patients previously submitted to callosotomy. *Seizure* **2009**, *18*, 588–592. [CrossRef] [PubMed]

29. Valentin, A.; Garcia Navarrete, E.; Chelvarajah, R.; Torres, C.; Navas, M.; Vico, L.; Torres, N.; Pastor, J.; Selway, R.; Sola, R.G.; et al. Deep brain stimulation of the centromedian thalamic nucleus for the treatment of generalized and frontal epilepsies. *Epilepsia* **2013**, *54*, 1823–1833. [CrossRef] [PubMed]

30. Valentin, A.; Nguyen, H.Q.; Skupenova, A.M.; Agirre-Arrizubieta, Z.; Jewell, S.; Mullatti, N.; Moran, N.F.; Richardson, M.P.; Selway, R.P.; Alarcon, G. Centromedian thalamic nuclei deep brain stimulation in refractory status epilepticus. *Brain Stimul.* **2012**, *5*, 594–598. [CrossRef] [PubMed]

31. Lehtimaki, K.; Langsjo, J.W.; Ollikainen, J.; Heinonen, H.; Mottonen, T.; Tahtinen, T.; Haapasalo, J.; Tenhunen, J.; Katisko, J.; Ohman, J.; et al. Successful management of super-refractory status epilepticus with thalamic deep brain stimulation. *Ann. Neurol.* **2017**, *81*, 142–146. [CrossRef] [PubMed]

32. Child, N.D.; Benarroch, E.E. Anterior nucleus of the thalamus: Functional organization and clinical implications. *Neurology* **2013**, *81*, 1869–1876. [CrossRef] [PubMed]

33. Hodaie, M.; Wennberg, R.A.; Dostrovsky, J.O.; Lozano, A.M. Chronic anterior thalamus stimulation for intractable epilepsy. *Epilepsia* **2002**, *43*, 603–608. [CrossRef] [PubMed]

34. Kerrigan, J.F.; Litt, B.; Fisher, R.S.; Cranstoun, S.; French, J.A.; Blum, D.E.; Dichter, M.; Shetter, A.; Baltuch, G.; Jaggi, J.; et al. Electrical stimulation of the anterior nucleus of the thalamus for the treatment of intractable epilepsy. *Epilepsia* **2004**, *45*, 346–354. [CrossRef] [PubMed]

35. Lim, S.N.; Lee, S.T.; Tsai, Y.T.; Chen, I.A.; Tu, P.H.; Chen, J.L.; Chang, H.W.; Su, Y.C.; Wu, T. Electrical stimulation of the anterior nucleus of the thalamus for intractable epilepsy: A long-term follow-up study. *Epilepsia* **2007**, *48*, 342–347. [CrossRef] [PubMed]

36. Osorio, I.; Overman, J.; Giftakis, J.; Wilkinson, S.B. High frequency thalamic stimulation for inoperable mesial temporal epilepsy. *Epilepsia* **2007**, *48*, 1561–1571. [CrossRef] [PubMed]

37. Oh, Y.S.; Kim, H.J.; Lee, K.J.; Kim, Y.I.; Lim, S.C.; Shon, Y.M. Cognitive improvement after long-term electrical stimulation of bilateral anterior thalamic nucleus in refractory epilepsy patients. *Seizure* **2012**, *21*, 183–187. [CrossRef] [PubMed]

38. Lehtimaki, K.; Mottonen, T.; Jarventausta, K.; Katisko, J.; Tahtinen, T.; Haapasalo, J.; Niskakangas, T.; Kiekara, T.; Ohman, J.; Peltola, J. Outcome based definition of the anterior thalamic deep brain stimulation target in refractory epilepsy. *Brain Stimul.* **2016**, *9*, 268–275. [CrossRef] [PubMed]

39. Piacentino, M.; Durisotti, C.; Garofalo, P.G.; Bonanni, P.; Volzone, A.; Ranzato, F.; Beggio, G. Anterior thalamic nucleus deep brain Stimulation (DBS) for drug-resistant complex partial seizures (CPS) with or without generalization: Long-term evaluation and predictive outcome. *Acta Neurochir.* **2015**, *157*, 1525–1532; discussion 1532. [CrossRef] [PubMed]

40. Lee, K.J.; Jang, K.S.; Shon, Y.M. Chronic deep brain stimulation of subthalamic and anterior thalamic nuclei for controlling refractory partial epilepsy. *Acta Neurochir. Suppl.* **2006**, *99*, 87–91. [PubMed]

41. Wiebe, S. Outcome measures in intractable epilepsy. *Adv. Neurol.* **2006**, *97*, 11–15. [PubMed]

42. Cohen-Gadol, A.A.; Wilhelmi, B.G.; Collignon, F.; White, J.B.; Britton, J.W.; Cambier, D.M.; Christianson, T.J.; Marsh, W.R.; Meyer, F.B.; Cascino, G.D. Long-term outcome of epilepsy surgery among 399 patients with nonlesional seizure foci including mesial temporal lobe sclerosis. *J. Neurosurg.* **2006**, *104*, 513–524. [CrossRef] [PubMed]

43. Wiebe, S.; Blume, W.T.; Girvin, J.P.; Eliasziw, M. Effectiveness and Efficiency of Surgery for Temporal Lobe Epilepsy Study Group. A randomized, controlled trial of surgery for temporal-lobe epilepsy. *N. Engl. J. Med.* **2001**, *345*, 311–318. [CrossRef] [PubMed]

44. Engel, J., Jr. Why is there still doubt to cut it out? *Epilepsy Curr.* **2013**, *13*, 198–204. [CrossRef] [PubMed]

45. Weiss, S.R.; Li, X.L.; Rosen, J.B.; Li, H.; Heynen, T.; Post, R.M. Quenching: Inhibition of development and expression of amygdala kindled seizures with low frequency stimulation. *Neuroreport* **1995**, *6*, 2171–2176. [CrossRef] [PubMed]

46. Weiss, S.R.; Eidsath, A.; Li, X.L.; Heynen, T.; Post, R.M. Quenching revisited: Low level direct current inhibits amygdala-kindled seizures. *Exp. Neurol.* **1998**, *154*, 185–192. [CrossRef] [PubMed]

47. Kile, K.B.; Tian, N.; Durand, D.M. Low frequency stimulation decreases seizure activity in a mutation model of epilepsy. *Epilepsia* **2010**, *51*, 1745–1753. [CrossRef] [PubMed]

48. Zhang, S.H.; Sun, H.L.; Fang, Q.; Zhong, K.; Wu, D.C.; Wang, S.; Chen, Z. Low-frequency stimulation of the hippocampal CA3 subfield is anti-epileptogenic and anti-ictogenic in rat amygdaloid kindling model of epilepsy. *Neurosci. Lett.* **2009**, *455*, 51–55. [CrossRef] [PubMed]

49. Wyckhuys, T.; De Smedt, T.; Claeys, P.; Raedt, R.; Waterschoot, L.; Vonck, K.; Van den Broecke, C.; Mabilde, C.; Leybaert, L.; Wadman, W.; et al. High frequency deep brain stimulation in the hippocampus modifies seizure characteristics in kindled rats. *Epilepsia* **2007**, *48*, 1543–1550. [CrossRef] [PubMed]

50. Wyckhuys, T.; Staelens, S.; Van Nieuwenhuyse, B.; Deleye, S.; Hallez, H.; Vonck, K.; Raedt, R.; Wadman, W.; Boon, P. Hippocampal deep brain stimulation induces decreased rCBF in the hippocampal formation of the rat. *Neuroimage* **2010**, *52*, 55–61. [CrossRef] [PubMed]

51. Velasco, A.L.; Velasco, M.; Velasco, F.; Menes, D.; Gordon, F.; Rocha, L.; Briones, M.; Marquez, I. Subacute and chronic electrical stimulation of the hippocampus on intractable temporal lobe seizures: Preliminary report. *Arch. Med. Res.* **2000**, *31*, 316–328. [CrossRef]

52. Tellez-Zenteno, J.F.; McLachlan, R.S.; Parrent, A.; Kubu, C.S.; Wiebe, S. Hippocampal electrical stimulation in mesial temporal lobe epilepsy. *Neurology* **2006**, *66*, 1490–1494. [CrossRef] [PubMed]

53. Velasco, A.L.; Velasco, F.; Velasco, M.; Trejo, D.; Castro, G.; Carrillo-Ruiz, J.D. Electrical stimulation of the hippocampal epileptic foci for seizure control: A double-blind, long-term follow-up study. *Epilepsia* **2007**, *48*, 1895–1903. [CrossRef] [PubMed]

54. McLachlan, R.S.; Pigott, S.; Tellez-Zenteno, J.F.; Wiebe, S.; Parrent, A. Bilateral hippocampal stimulation for intractable temporal lobe epilepsy: Impact on seizures and memory. *Epilepsia* **2010**, *51*, 304–307. [CrossRef] [PubMed]

55. Wiebe, S.; Kiss, Z.; Ahmed, N. Medical vs. electrical therapy for mesial temporal lobe epilepsy: A multicenter randomized trial. *Epilepsy Curr.* **2013**, *13* (Suppl. 1), 288.

56. Cukiert, A.; Cukiert, C.M.; Burattini, J.A.; Mariani, P.P.; Bezerra, D.F. Seizure outcome after hippocampal deep brain stimulation in patients with refractory temporal lobe epilepsy: A prospective, controlled, randomized, double-blind study. *Epilepsia* **2017**, *58*, 1728–1733. [CrossRef] [PubMed]

57. Velasco, A.L.; Velasco, F.; Velasco, M.; Jimenez, F.; Carrillo-Ruiz, J.D.; Castro, G. The role of neuromodulation of the hippocampus in the treatment of intractable complex partial seizures of the temporal lobe. *Acta Neurochir. Suppl.* **2007**, *97*, 329–332. [PubMed]

58. Cukiert, A.; Cukiert, C.M.; Argentoni-Baldochi, M.; Baise, C.; Forster, C.R.; Mello, V.A.; Burattini, J.A.; Lima, A.M. Intraoperative neurophysiological responses in epileptic patients submitted to hippocampal and thalamic deep brain stimulation. *Seizure* **2011**, *20*, 748–753. [CrossRef] [PubMed]

59. Cukiert, A.; Cukiert, C.M.; Burattini, J.A.; Lima, A.M. Seizure outcome after hippocampal deep brain stimulation in a prospective cohort of patients with refractory temporal lobe epilepsy. *Seizure* **2014**, *23*, 6–9. [CrossRef] [PubMed]

60. Sturm, V.; Lenartz, D.; Koulousakis, A.; Treuer, H.; Herholz, K.; Klein, J.C.; Klosterkotter, J. The nucleus accumbens: A target for deep brain stimulation in obsessive-compulsive- and anxiety-disorders. *J. Chem. Neuroanat.* **2003**, *26*, 293–299. [CrossRef] [PubMed]

61. Lothman, E.W.; Hatlelid, J.M.; Zorumski, C.F. Functional mapping of limbic seizures originating in the hippocampus: A combined 2-deoxyglucose and electrophysiologic study. *Brain Res.* **1985**, *360*, 92–100. [CrossRef]

62. Ma, J.; Leung, L.S. Kindled seizure in the prefrontal cortex activated behavioral hyperactivity and increase in accumbens gamma oscillations through the hippocampus. *Behav. Brain Res.* **2010**, *206*, 68–77. [CrossRef] [PubMed]

63. Kowski, A.B.; Voges, J.; Heinze, H.J.; Oltmanns, F.; Holtkamp, M.; Schmitt, F.C. Nucleus accumbens stimulation in partial epilepsy—A randomized controlled case series. *Epilepsia* **2015**, *56*, e78–e82. [CrossRef] [PubMed]

64. Klinger, N.V.; Mittal, S. Clinical efficacy of deep brain stimulation for the treatment of medically refractory epilepsy. *Clin. Neurol. Neurosurg.* **2016**, *140*, 11–25. [CrossRef] [PubMed]

65. Chabardes, S.; Kahane, P.; Minotti, L.; Koudsie, A.; Hirsch, E.; Benabid, A.L. Deep brain stimulation in epilepsy with particular reference to the subthalamic nucleus. *Epileptic Disord.* **2002**, *4* (Suppl. 3), S83–S93. [PubMed]

66. Wille, C.; Steinhoff, B.J.; Altenmuller, D.M.; Staack, A.M.; Bilic, S.; Nikkhah, G.; Vesper, J. Chronic high-frequency deep-brain stimulation in progressive myoclonic epilepsy in adulthood—Report of five cases. *Epilepsia* **2011**, *52*, 489–496. [CrossRef] [PubMed]

67. Capecci, M.; Ricciuti, R.A.; Ortenzi, A.; Paggi, A.; Durazzi, V.; Rychlicki, F.; Provinciali, L.; Scerrati, M.; Ceravolo, M.G. Chronic bilateral subthalamic stimulation after anterior callosotomy in drug-resistant epilepsy: Long-term clinical and functional outcome of two cases. *Epilepsy Res.* **2012**, *98*, 135–139. [CrossRef] [PubMed]

68. Chkhenkeli, S.A.; Chkhenkeli, I.S. Effects of therapeutic stimulation of nucleus caudatus on epileptic electrical activity of brain in patients with intractable epilepsy. *Stereotact. Funct. Neurosurg.* **1997**, *69*, 221–224. [CrossRef] [PubMed]

69. Chkhenkeli, S.A.; Sramka, M.; Lortkipanidze, G.S.; Rakviashvili, T.N.; Bregvadze, E.; Magalashvili, G.E.; Gagoshidze, T.; Chkhenkeli, I.S. Electrophysiological effects and clinical results of direct brain stimulation for intractable epilepsy. *Clin. Neurol. Neurosurg.* **2004**, *106*, 318–329. [CrossRef] [PubMed]

70. Stagg, C.J.; Best, J.G.; Stephenson, M.C.; O'Shea, J.; Wylezinska, M.; Kincses, Z.T.; Morris, P.G.; Matthews, P.M.; Johansen-Berg, H. Polarity-sensitive modulation of cortical neurotransmitters by transcranial stimulation. *J. Neurosci.* **2009**, *29*, 5202–5206. [CrossRef] [PubMed]

71. Cogiamanian, F.; Vergari, M.; Pulecchi, F.; Marceglia, S.; Priori, A. Effect of spinal transcutaneous direct current stimulation on somatosensory evoked potentials in humans. *Clin. Neurophysiol.* **2008**, *119*, 2636–2640. [CrossRef] [PubMed]

72. San-Juan, D.; Lopez, D.A.E.; Gregorio, R.V.; Trenado, C.; Aragon, M.F.G.; Morales-Quezada, L.; Ruiz, A.H.; Hernandez-Gonzalez, F.; Alcaraz-Guzman, A.; Anschel, D.J.; et al. Transcranial Direct Current Stimulation in Mesial Temporal Lobe Epilepsy and Hippocampal Sclerosis. *Brain Stimul.* **2017**, *10*, 28–35. [CrossRef] [PubMed]

73. Schweitzer, A.; Wright, S. Effects on the knee jerk of stimulation of the central end of the vagus and of various changes in the circulation and respiration. *J. Physiol.* **1937**, *88*, 459–475. [CrossRef] [PubMed]

74. Rutecki, P. Anatomical, Physiological, and Theoretical Basis for the Antiepileptic Effect of Vagus Nerve-Stimulation. *Epilepsia* **1990**, *31*, S1–S6. [CrossRef] [PubMed]

75. Groves, D.A.; Brown, V.J. Vagal nerve stimulation: A review of its applications and potential mechanisms that mediate its clinical effects. *Neurosci. Biobehav. Rev.* **2005**, *29*, 493–500. [CrossRef] [PubMed]

76. Vonck, K.; De Herdt, V.; Boon, P. Vagal nerve stimulation—A 15-year survey of an established treatment modality in epilepsy surgery. *Adv. Tech. Stand. Neurosurg.* **2009**, *34*, 111–146. [PubMed]

77. Chae, J.H.; Nahas, Z.; Lomarev, M.; Denslow, S.; Lorberbaum, J.P.; Bohning, D.E.; George, M.S. A review of functional neuroimaging studies of vagus nerve stimulation (VNS). *J. Psychiatr. Res.* **2003**, *37*, 443–455. [CrossRef]

78. Benmenachem, E.; Manonespaillat, R.; Ristanovic, R.; Wilder, B.J.; Stefan, H.; Mirza, W.; Tarver, W.B.; Wernicke, J.F.; Augustinsson, L.; Barolat, G.; et al. Vagus Nerve-Stimulation for Treatment of Partial Seizures: 1. A Controlled-Study of Effect on Seizures. *Epilepsia* **1994**, *35*, 616–626. [CrossRef]

79. Handforth, A.; DeGiorgio, C.M.; Schachter, S.C.; Uthman, B.M.; Naritoku, D.K.; Tecoma, E.S.; Henry, T.R.; Collins, S.D.; Vaughn, B.V.; Gilmartin, R.C.; et al. Vagus nerve stimulation therapy for partial-onset seizures—A randomized active-control trial. *Neurology* **1998**, *51*, 48–55. [CrossRef] [PubMed]

80. Klinkenberg, S.; Aalbers, M.W.; Vles, J.S.H.; Cornips, E.M.J.; Rijkers, K.; Leenen, L.; Kessels, F.G.H.; Aldenkamp, A.P.; Majoie, M. Vagus nerve stimulation in children with intractable epilepsy: A randomized controlled trial. *Dev. Med. Child Neurol.* **2012**, *54*, 855–861. [CrossRef] [PubMed]

81. Ryvlin, P.; Gilliam, F.G.; Nguyen, D.K.; Colicchio, G.; Iudice, A.; Tinuper, P.; Zamponi, N.; Aguglia, U.; Wagner, L.; Minotti, L. The long-term effect of vagus nerve stimulation on quality of life in patients with pharmacoresistant focal epilepsy: The PuLsE (Open Prospective Randomized Long-term Effectiveness) trial. *Epilepsia* **2014**, *55*, 893–900. [CrossRef] [PubMed]

82. Aihua, L.; Lu, S.; Liping, L.; Xiuru, W.; Hua, L.; Yuping, W. A controlled trial of transcutaneous vagus nerve stimulation for the treatment of pharmacoresistant epilepsy. *Epilepsy Behav.* **2014**, *39*, 105–110. [CrossRef] [PubMed]

83. Magdaleno-Madrigal, V.M.; Valdes-Cruz, A.; Martinez-Vargas, D.; Martinez, A.; Almazan, S.; Fernandez-Mas, R.; Fernandez-Guardiola, A. Effect of electrical stimulation of the nucleus of the solitary tract on the development of electrical amygdaloid kindling in the cat. *Epilepsia* **2002**, *43*, 964–969. [CrossRef] [PubMed]

84. Neuman, R.S. Suppression of Penicillin-Induced Focal Epileptiform Activity by Locus-Ceruleus Stimulation—Mediation by an Alpha-1-Adrenoceptor. *Epilepsia* **1986**, *27*, 359–366. [CrossRef] [PubMed]

85. DeGiorgio, C.M.; Shewmon, D.A.; Whitehurst, T. Trigeminal nerve stimulation for epilepsy. *Neurology* **2003**, *61*, 421–422. [CrossRef] [PubMed]

86. Fanselow, E.E.; Reid, A.P.; Nicolelis, M.A.L. Reduction of pentylenetetrazole-induced seizure activity in awake rats by seizure-triggered trigeminal nerve stimulation. *J. Neurosci.* **2000**, *20*, 8160–8168. [CrossRef] [PubMed]

87. DeGiorgio, C.M.; Soss, J.; Cook, I.A.; Markovic, D.; Gornbein, J.; Oviedo, D.M.S.; Oviedo, S.; Gordon, S.; Corralle-Leyva, G.; Kealey, C.P.; et al. Randomized controlled trial of trigeminal nerve stimulation for drug-resistant epilepsy. *Neurology* **2013**, *80*, 786–791. [CrossRef] [PubMed]

88. Faught, E.; Tatum, W. Trigeminal stimulation A superhighway to the brain? *Neurology* **2013**, *80*, 780–781. [CrossRef] [PubMed]

89. Kimiskidis, V.K. Transcranial Magnetic Stimulation for Drug-Resistant Epilepsies: Rationale and Clinical Experience. *Eur. Neurol.* **2010**, *63*, 205–210. [CrossRef] [PubMed]

90. Reithler, J.; Peters, J.C.; Sack, A.T. Multimodal transcranial magnetic stimulation: Using concurrent neuroimaging to reveal the neural network dynamics of noninvasive brain stimulation. *Prog. Neurobiol.* **2011**, *94*, 149–165. [CrossRef] [PubMed]

91. Huang, Y.Z.; Edwards, M.J.; Rounis, E.; Bhatia, K.P.; Rothwell, J.C. Theta burst stimulation of the human motor cortex. *Neuron* **2005**, *45*, 201–206. [CrossRef] [PubMed]

92. Fregni, F.; Otachi, P.T.M.; do Valle, A.; Boggio, P.S.; Thut, G.; Rigonatti, S.P.; Pascual-Leone, A.; Valente, K.D. A randomized clinical trial of repetitive transcranial magnetic stimulation in patients with refractory epilepsy. *Ann. Neurol.* **2006**, *60*, 447–455. [CrossRef] [PubMed]

93. Sun, W.; Mao, W.; Meng, X.H.; Wang, D.Q.; Qiao, L.; Tao, W.; Li, L.P.; Jia, X.Y.; Han, C.Y.; Fu, M.M.; et al. Low-frequency repetitive transcranial magnetic stimulation for the treatment of refractory partial epilepsy: A controlled clinical study. *Epilepsia* **2012**, *53*, 1782–1789. [CrossRef] [PubMed]

94. Wang, X.X.; Yang, D.B.; Wang, S.X.; Zhao, X.Q.; Zhang, L.L.; Chen, Z.Q.; Sun, X.R. Effects of low-frequency repetitive transcranial magnetic stimulation on electroencephalogram and seizure frequency in 15 patients with temporal lobe epilepsy following dipole source localization. *Neural Regen. Res.* **2008**, *3*, 1257–1260.

95. Cantello, R.; Rossi, S.; Varrasi, C.; Ulivelli, M.; Civardi, C.; Bartalini, S.; Vatti, G.; Cincotta, M.; Borgheresi, A.; Zaccara, G.; et al. Slow repetitive TMS for drug-resistant epilepsy: Clinical and EEG findings of a placebo-controlled trial. *Epilepsia* **2007**, *48*, 366–374. [CrossRef] [PubMed]

96. Morrell, M.J. Responsive cortical stimulation for the treatment of medically intractable partial epilepsy. *Neurology* **2011**, *77*, 1295–1304. [CrossRef] [PubMed]

97. Heck, C.N.; King-Stephens, D.; Massey, A.D.; Nair, D.R.; Jobst, B.C.; Barkley, G.L.; Salanova, V.; Cole, A.J.; Smith, M.C.; Gwinn, R.P.; et al. Two-year seizure reduction in adults with medically intractable partial onset epilepsy treated with responsive neurostimulation: Final results of the RNS System Pivotal trial. *Epilepsia* **2014**, *55*, 432–441. [CrossRef] [PubMed]

98. Geller, E.B.; Skarpaas, T.L.; Gross, R.E.; Goodman, R.R.; Barkley, G.L.; Bazil, C.W.; Berg, M.J.; Bergey, G.K.; Cash, S.S.; Cole, A.J.; et al. Brain-responsive neurostimulation in patients with medically intractable mesial temporal lobe epilepsy. *Epilepsia* **2017**, *58*, 994–1004. [CrossRef] [PubMed]

99. Jobst, B.C.; Kapur, R.; Barkley, G.L.; Bazil, C.W.; Berg, M.J.; Bergey, G.K.; Boggs, J.G.; Cash, S.S.; Cole, A.J.; Duchowny, M.S.; et al. Brain-responsive neurostimulation in patients with medically intractable seizures arising from eloquent and other neocortical areas. *Epilepsia* **2017**, *58*, 1005–1014. [CrossRef] [PubMed]

100. Child, N.D.; Stead, M.; Wirrell, E.C.; Nickels, K.C.; Wetjen, N.M.; Lee, K.H.; Klassen, B.T. Chronic subthreshold subdural cortical stimulation for the treatment of focal epilepsy originating from eloquent cortex. *Epilepsia* **2014**, *55*, e18–e21. [CrossRef] [PubMed]

101. Valentin, A.; Ughratdar, I.; Cheserem, B.; Morris, R.; Selway, R.; Alarcon, G. Epilepsia partialis continua responsive to neocortical electrical stimulation. *Epilepsia* **2015**, *56*, e104–e109. [CrossRef] [PubMed]

102. Lundstrom, B.N.; Van Gompel, J.; Britton, J.; Nickels, K.; Wetjen, N.; Worrell, G.; Stead, M. Chronic Subthreshold Cortical Stimulation to Treat Focal Epilepsy. *JAMA Neurol.* **2016**, *73*, 1370–1372. [CrossRef] [PubMed]

MDPI

St. Alban-Anlage 66

4052 Basel

Switzerland

Tel. +41 61 683 77 34

Fax +41 61 302 89 18

www.mdpi.com

Brain Sciences Editorial Office

E-mail: brainsci@mdpi.com

www.mdpi.com/journal/brainsci